Teaching Probability

Jenny Gage   David Spiegelhalter

Cambridge Mathematics

CAMBRIDGE
UNIVERSITY PRESS

Shaftesbury Road, Cambridge CB2 8EA, United Kingdom

One Liberty Plaza, 20th Floor, New York, NY 10006, USA

477 Williamstown Road, Port Melbourne, VIC 3207, Australia

314–321, 3rd Floor, Plot 3, Splendor Forum, Jasola District Centre, New Delhi – 110025, India

103 Penang Road, #05–06/07, Visioncrest Commercial, Singapore 238467

Cambridge University Press is part of the University of Cambridge.

It furthers the University's mission by disseminating knowledge in the pursuit of education, learning and research at the highest international levels of excellence.

www.cambridge.org
Information on this title: www.cambridge.org/9781316605899

© Cambridge University Press & Assessment 2018

First published 2016

20 19 18 17 16 15 14 13 12 11 10 9 8 7 6 5 4 3 2

A catalogue record for this publication is available from the British Library

ISBN   978-1-316-60589-9   Paperback

Additional resources for this publication at www.teachingprobability.org

Cambridge University Press has no responsibility for the persistence or accuracy of URLs for external or third-party internet websites referred to in this publication, and does not guarantee that any content on such websites is, or will remain, accurate or appropriate. Information regarding prices, travel timetables, and other factual information given in this work is correct at the time of first printing but Cambridge University Press does not guarantee the accuracy of such information thereafter.

....

Contents

Acknowledgements

We would like to thank the African Institute for Mathematical Sciences Schools Enrichment Centre (AIMSSEC) for their contribution to the development of the approach in this book. Jenny had taught probability to many secondary school students before teachers in South Africa helped her to realise she was starting in the wrong place, and that some of her tried and tested approaches didn't work in a different culture. One big problem was the use of dice. Not only can it be culturally inappropriate (because of associations with gambling), but some found it difficult to see the numbers on a die as labels which can be used to signify something else – it wasn't necessarily obvious that, for instance, a 1 or a 2 might mean a goal (but only one goal) to the Beavers, while 3, 4, 5 and 6 meant one goal (only) to the Raccoons. This raised the question of whether such difficulties might be more widespread. The spinners we use in this book provide a perfect way around this, and we are very grateful to Claire Blackman, now at the University of Cape Town but then working with AIMSSEC, for this idea.

We are very grateful to The King's School, Ely, for their willingness to allow us to provide lesson materials, observe lessons and take photos trialling these versions of the classroom resources. We would like to thank Nadia Baker and her students for their help and enthusiasm. Not only did Nadia willingly give up time for these trials, she also contributed to the early development of our approach when she worked for the Millennium Mathematics Project and shared an office, and many ideas, with Jenny. The serendipity of that and her own work with David on the Probability Roadshow are part of the genesis of this book.

We are thankful to the many teachers in Cambridgeshire and Buckinghamshire who tested early versions of the resources in their classrooms and whose responses encouraged us to continue.

We would also like to acknowledge the inspiration of Professor Gerd Gigerenzer, who has tirelessly promoted the use of 'natural', or expected, frequencies as a way of improving the public understanding of probability. Finally, we are deeply grateful to David Harding of Winton Capital Management for both funding David's post and providing continuing support and enthusiasm.

International studies of mathematics curricula reveal that the teaching of probability takes up relatively little time within the mathematics timetable. This is simultaneously disappointing and unsurprising. Disappointing because probability and the associated understanding of risk should contribute hugely to the development of well-informed citizens. In an era when we are surrounded and bombarded by media articles and public figures offering often dubious statistics, it is important that we prepare the next generation to be able to evaluate these critically and make informed judgements. And it's unsurprising because, historically, probability is notoriously a tricky topic to teach and difficult to learn, and often teachers are reluctant to spend longer on it than is absolutely necessary. Much of students' experience, therefore, is squeezed into a short period of time in which they skim the surface of probability and simply learn to answer high-stakes assessment questions by rote.

But things are changing …

All over the world, policy makers are recognising that the point of learning mathematics is to solve problems, whether of the real-world variety or within the mathematics domain itself. Whilst knowledge and facts are important, being able to reason with them and apply them in non-routine situations requires teaching for deeper understanding. And that's what this book is all about.

The NRICH project (*www.nrich.maths.org*) has a long tradition of working with experts in the field of mathematics and, being situated in the Faculty of Mathematics within the University of Cambridge, there are plenty of highly qualified mathematicians at hand. When David Spiegelhalter, the Winton Professor for the Public Understanding of Risk, showed an interest in designing innovative rich tasks to support the teaching of probability, we jumped at the opportunity – who wouldn't! Jenny Gage was a member of the NRICH team at that time and so began a fruitful partnership, the result of which is this book.

Whilst working with NRICH, Jenny and David devised activities based on evidence from risk education, a field in which David in particular has been very involved. Whilst acknowledging that there are several aspects of probability that make it less intuitive than might be imagined, they reframe the progression in conceptual understanding by relating traditional probability exercises to situations that can be modelled using expected frequencies. They suggest that when students are presented with different representations of expected frequency – frequency trees, Venn diagrams and two-way tables – they notice different aspects of the task. These varied experiences then ensure secure foundations for later work with the more abstract representations and calculations.

Part 1 sets the scene for the rest of the book. It describes the evidence behind the design principles, and explores and explains the progression, illustrated with classroom dialogues. The part finishes with a suggested curriculum, setting out the content with exemplar questions and a helpful commentary. All of this is enlarged upon in Part 2, where the reader is introduced to rich classroom activities. Extensive support notes and exemplar solutions are also provided, along with students' work from extensive trialling – all of which give the activities added credibility. Both here and in Part 4 there are links to online resources; many of these come from projects that David and Jenny have been involved with, such as Motivate, Plus, Understanding Uncertainty, and the original activities on the NRICH site that were the stimulus for this book.

Part 3 is a collection of typical assessment questions, grouped by predominant content, with a range of alternative solutions and commentary. Part 4 comprises additional activities which could be used as enrichment in the classroom but, for this reader, served to explain and challenge some of the popular questions and misconceptions involving probability. I particularly enjoyed the references to other eminent writers. As with the rest of the book, this part is presented in the authors' characteristically accessible style, with an eye for the ridiculous or humorous, and treats the reader as an intelligent partner in a joint enterprise.

You may have noticed that this volume carries the Cambridge Mathematics logo. I'm delighted it is our first publication and so is unique!

Cambridge Mathematics is an initiative of four partner departments of the University of Cambridge – the Faculties of Mathematics and Education, Cambridge Assessment and Cambridge University Press. Together we are committed to championing and securing a world-class mathematics education for all students from 5 to 19 years old, applicable to both national and international contexts and based on evidence from research and practice. We realise this is a hugely ambitious aim but think the time is right to call on our considerable institutional resources – and the support of our national and international colleagues – to develop a coherent, transparent, evidence-based vision that will make a major contribution to mathematics education internationally.

The first piece of work is in developing a Framework that will be the spine for the whole initiative. This is a mapping of all the mathematics which we think should be offered to students from age 5 to 19, no matter what their purpose is in studying mathematics. In designing this we need to equip students for futures which currently we can't predict – in all occupations and walks of life. Our vision is to devise a model of mathematics education that will take into account the different and enduring needs of students in a changing world and contribute to students' personal, societal and economic wellbeing. It will emphasise the richness and power of maths, will encourage continued study of the subject and will be recognised worldwide as innovative and rigorous in approach.

We are fortunate in being able to call on experts from across the world to advise and support us in this endeavour. In addition to the Framework, we will publish resources, both for the classroom and for teachers. This book, written by experts and in partnership with Cambridge Mathematics, therefore fits our aims, principles and plans. We're delighted!

Lynne McClure
Director, Cambridge Mathematics
Cambridge, Summer 2016

Aims and Principles of Cambridge Mathematics

Cambridge Mathematics is a long-term programme of developments and by 2020 we aim to have made considerable progress towards our seven subsidiary aims:

- to champion and secure access to a high-quality maths education for all learners
- to collaborate, using our position in maths education to show leadership and develop an authoritative voice
- to develop a coherent Cambridge Mathematics Framework for all ages and types of learner, with a strong distinctive approach led by academics and educationalists and supported by a strong research base
- to develop and make available world-class teaching and learning materials
- to support an infrastructure that enhances the quality of teacher education and continuing professional development
- to develop assessments that support the development of powerful mathematical reasoning
- to develop an approach that is recognised and valued by parents, young people, teachers, institutions and governments.

The four principles underpinning our work are:

- access for all – we will champion access for all students
- evidence based – we will use rigorous research to determine the most effective ways of working that will improve outcomes
- collaboration and consultation – we will work and consult with partners in mathematics education both nationally and internationally, and in the public and private sectors
- coherent and integrated – the four integrated elements of Cambridge Mathematics are:
  - the Cambridge Mathematics Framework, the content spine to which the other elements will link
  - resources, both paper-based and electronic
  - a coherent formative and summative assessment offer
  - a professional development framework encompassing both subject and pedagogical knowledge.

www.cambridgemaths.org

Additional resources for this book are available at:

www.teachingprobability.org

This website includes spinners, worksheets, spreadsheets and links to further resources.

Chapter

1

Introduction

1.1 What this book is about

Probability is the branch of mathematics that deals with randomness, chance, unpredictability and risk. These are vital issues for everyone in society – we all need to make decisions in the face of uncertainty. And yet the public's ability to reason with probability is dismally poor. It is also not generally a popular part of the school mathematics syllabus, either for teachers or for students.

However, researchers in risk communication have shown that changing the way in which probability is represented can dramatically improve people's ability to carry out quite complex tasks. Instead of talking about chance or probability in terms of a decimal, percentage or fraction, we look at the **expected frequency** of events in a group of cases. For example, when discussing the risk of a future heart attack or stroke with a patient, medical students are now taught not to say 'a 16% chance', but instead to say 'out of 100 patients like you, we would expect 16 to have a heart attack or stroke in the next ten years'.

This may seem a trivial change, but it has strong implications for the way in which probability is taught in schools. Our aim is not only to enable students to answer the type of probability questions set in examinations, but also to help them handle uncertainty in the world beyond the classroom.

1.2 Probability is important

Life is uncertain. None of us knows what is going to happen so, unless we are prepared to resign ourselves to fate, it seems a good idea to be able to reason about uncertainty. Whether we are deciding about medical treatments, choosing investments, buying insurance, playing games or undertaking a risky activity, we want to be able to weigh up the options in terms of the chances, and consequences, of the good and bad things that might happen. Probability is also the basis for methods used in forecasting the economy, the weather or epidemics, as well as underlying much of physics. When a scientific discovery such as the Higgs boson is claimed, the degree of certainty, or confidence, is expressed in terms of probability since probability forms the basis for statistical inference.

Even without its clear practical importance, probability can be fascinating in its own right and provides a starting point for a wealth of challenging problems and games, as well as (in some cultures) gambling.

There is also the association between probability and fairness. The idea of 'casting lots' as a fair way of making decisions or allocating goods is ancient, and children are sensitised to the link between pure randomness and fairness from an early age. Technically, we can identify 'fairness' with the idea that each individual has the same expected gain, which leads us back to the need to understand probability.

1.3 What is probability anyway?

So probability is important, and its applications are all around us. Why, then, do people find it so unintuitive and difficult? Well, after years of working in this area, we have finally concluded that this is because … probability *is* unintuitive and difficult.

People's understanding is not helped by the lack of clarity about what probability actually *is*. We have scales for weight, rulers for length, clocks for time, but where is the probability-meter? Probability, like value, is not directly and objectively measurable. What is worse, philosophers of science have been unable to come up with an agreed definition for probability, and so it is impossible to specify exactly what it is.

Some popular options for the definition of the probability of an event include:

a   **Symmetry**: 'The number of outcomes favouring the event, divided by the total number of outcomes, assuming the outcomes are all equally likely.' This is the definition usually taught in school as **theoretical probability**, but it is rather circular as it depends on 'equally likely' being defined. And, it can only be used in nicely balanced situations such as dice, cards or lottery tickets, or when, to use a classic example, picking a coloured sock at random from a drawer. It does not apply, for example, to the probability that you will have a heart attack or stroke in the next ten years.

b   **Frequency**: 'The proportion of times, in the long run of identical circumstances, that the event occurs.' This is the idea of an observed **relative frequency** tending to a true probability after sufficient repetitions. This can be fine for situations where there are lots of repeats, but does not seem applicable to unique situations, such as *your* risk of a heart attack.

c   **Subjective**: 'My personal confidence that an event will occur, expressed as a number between 0 and 1. When the event either occurs or not, my assessment will be rewarded or penalised according to an appropriate 'scoring rule'.' This definition is one way of formalising the idea that probabilities are purely personal judgements based on available evidence and assumptions.

There are other proposals for understanding probability. Some have suggested it measures an underlying **propensity** for an event to happen – but what is your propensity to have a heart attack or stroke? Or, more imaginatively, we could think of probability as the proportion of **possible futures** in which the event occurs.

For us, the crucial lessons from all this philosophy are:

a   We should not claim to have *the* definition of probability – it is
a 'virtual' quantity and perhaps best considered in terms of different
metaphors depending on circumstances.

b   Probabilities are almost inevitably based on judgements and assumptions
such as random sampling. They cannot be said to objectively exist,
except perhaps for sub-atomic, determined probabilities.

c   It is important to emphasise that, despite all these philosophical
debates, the mathematics of probability is not controversial.

In this book we primarily adopt a rather hybrid metaphor for probability,
based on the expected proportion of times that something will happen
in similar circumstances. This is essentially a frequency interpretation of a
subjective judgement. Using this idea, we show that complex probability
calculations can become remarkably clear.

1.4 People find probability tricky

The language of probability is complex and invites misunderstanding.
Suppose you are assessed to have a 16% probability of a heart attack or
stroke in the next ten years. Verbal terms are ambiguous and dependent
on context and viewpoint: we might personally think that 16% meant
this was a 'fairly unlikely' event, although from a medical point of view
this could be considered as 'high risk' and perhaps a cholesterol-lowering
drug, such as a statin, would be recommended.

Alternatively, we might describe this as around a 1 in 6 chance, but modern
advice in risk communication explicitly recommends against this type of
expression. For example, a recent population survey by telephone [1] asked:

Which of the following numbers represents the biggest risk of getting a disease:
1 in 100, 1 in 1000, or 1 in 10?

In Germany, 28% of responses were incorrect, and in the USA 25% were
wrong. The crucial issue is that *larger* numbers are used to communicate
smaller risks, so a difficult inversion must be done. This is one reason why
flood-risk maps expressed in terms of '1 in 100 year events' are difficult
to read and potentially misleading.

The media are also fond of reporting **relative risks**. For example,
an American direct-to-consumer advert for a statin to reduce cholesterol
declared in large font that there is a '36% reduction' in the risk of
heart attack. In very much smaller font it clarifies that this is a reduction,
in percentage point terms, from 3% to 2% over five years. This is a
reduction of 1 percentage point in the absolute risk, and so 100 such
people would have to take the drug every day for five years to prevent
one heart attack. This does not sound so impressive.

People are also not very good at handling relative risks. The same recent survey asked:

If person A's chance of getting a disease is 1 in 100 in 10 years and person B's risk is double that of A, what is B's risk?

¹ *Spoiler alert: the answer is that B's risk is 2 in 100 in 10 years.*

46% of respondents in Germany and 43% in the USA could not answer correctly.[1]

Even if the meaning of the probability statement is clear, there are numerous examples of the trouble people have with even fairly basic probabilistic reasoning.

In 2012, for example, 97 British Members of Parliament were asked:

If you spin a coin twice, what is the probability of getting two Heads?

Only 40% were able to answer correctly [2].

Because a question concerning probability is generally very easy to state (although sometimes ambiguous), people feel the answer should be intuitive. It rarely is. Even when trained, people can find it difficult to match the formal technique to the problem.

The only gut feeling we have about probability is not to trust our gut feelings.

1.5 There's a way to make probability less tricky

Our approach in this book is based on the research of psychologists into the effect different representations have on people's ability to reason with probabilities. The German psychologist, Gerd Gigerenzer, has popularised the idea of 'natural frequencies', which we call 'expected frequencies'. Extensive research [3–5] has shown this helps to prevent confusion and make probability calculations easier and more intuitive.

We have already revealed the basic idea: instead of saying 'the probability of X is 0.20 (or 20%)', we would say 'out of 100 situations like this, we would expect X to occur 20 times'. 'Is that all?', we hear you cry, but this simple re-expression can have a deep impact.

The first point is that it helps clarify what the probability means. When we hear the phrase 'the probability it will rain tomorrow is 30%', what does it mean? That it will rain 30% of the time? That it will rain over 30% of the area? In fact it means that out of 100 computer forecasts in situations like this, we expect rain in 30 of them. By clearly stating the denominator, or **reference class**, ambiguity is avoided.

An explicit reference class might have avoided some other journalistic mistakes, such as when it was reported that '35% of bikers have serious road accidents', when the real statistic was that 35% of serious road accidents involve motorcyclists.

Any proportion has a numerator and a denominator. Here the numerator is easy: bikers who have accidents. The problem comes with the denominator: in this situation it is not 'all bikers', it is 'all serious accidents'.

Or take the extraordinary headline that in Britain '30% of sex involves under 16s', when the actual claim was that 30% of under-16s have sex. Again the numerator is clearly 'under 16s having sex', but the journalist has taken the denominator as 'all sex' rather than the correct 'all under 16s'.

The crucial question is always to ask 'Out of what?', and then make this reference class explicit.

Even if you are using expected frequencies, such as '20 out of 100', to express risk you must keep in mind that the mathematically equivalent '200 out of 1000' suggests to many people a bigger risk, as the numerator is larger. This is known as **ratio bias**. The following example illustrates the difficulty students (and others) can have [6].

PISA 2003 included the following question.

Consider two boxes A and B. Box A contains three marbles, of which one is white and two are black. Box B contains 7 marbles, of which two are white and five are black. You have to draw a marble from one of the boxes with your eyes covered. From which box should you draw if you want a white marble?

The PISA 2003 Report commented that only 27% of the German school students obtained the correct answer.[2]

---

[2] Box A should be chosen since the chance of winning is $\frac{1}{3}$, which is larger than the chance with Box B, $\frac{2}{7}$. We recommend the following thought experiment to clarify the issue: in 21 replications of the experiment, how many times would you expect to win if you always chose Box A, or always chose Box B? You would expect 7 wins with Box A, and 6 wins with Box B.

Once again, this shows people being misled by focusing on the numerator, where Box B has the larger number of white marbles, 2. Focusing on the fraction (rather than the number) that are white, which is $\frac{1}{3}$ for Box A compared to $\frac{2}{7}$ for Box B, gives the correct answer – Box A. The extreme version of this bias, in which the denominator is ignored completely, is known as **denominator neglect**; the media do this every time they concentrate on a single accident without, for example, mentioning the millions of children who go to school safely each day [7].

Research has shown that, by using expected frequencies, people find it easier to carry out non-intuitive conditional probability calculations. Take a recent newspaper headline saying that eating 50 grams of processed meat each day (e.g. a bacon sandwich) is associated with a 20% increased risk of pancreatic cancer. It turns out that this very serious disease affects only 1 in 80 people. So we want to calculate a 20% increase on a 1 in 80 chance, which is tricky to do.

However, if we imagine 400 people who have an average breakfast each day, we can easily calculate that '1 in 80' means we would expect 5 out of the 400 to get pancreatic cancer. If 400 different people all stuff themselves with a greasy bacon sandwich every day of their lives, this 5 would increase by 20% to 6. This is actually a 1 in 400, or 0.25%, increase in absolute risk, which does not seem so important. Note the trick is in identifying 400 as the denominator that will lead to precisely one extra case due to excessive bacon consumption.

As mentioned previously, expected frequency is the standard format taught to medical students for risk communication, and is used extensively in public dialogue. In the advice leaflets for breast cancer screening in the UK, for example, the benefits and risks of screening are communicated in terms of what it means for 200 women being screened for 20 years: we would expect 1 woman to have her early death from breast cancer prevented by screening, at a cost of 3 women with non-threatening cancers being unnecessarily treated [8]. The key idea is that, through using whole numbers, we can think of the information as representing simple summaries of many possible experiences.

Expected frequencies can also be used to answer advanced conditional probability problems of the following classic type.

Suppose a screening test for doping in sports is claimed to be '95% accurate', meaning that 95% of dopers, and 95% of non-dopers, will be correctly classified. Assume 1 in 50 athletes are truly doping at any time. If an athlete tests positive, what is the probability that they are truly doping?

The way to answer such questions is to think of what we would expect to happen for, say, every 1000 tests conducted. Out of these, 1 in 50 (20) will be true dopers, of which 95% (19) will be correctly detected. But of the 980 non-dopers, 5% (49) will incorrectly test positive. That means a total of 68 positive tests, of which 19 are true dopers. So the probability that someone who tests positive is truly doping is $\frac{19}{68}$ = 28%. So, among the positive tests, the **false–positive** results greatly outnumber the correct detections by around 2.5 to 1.

If you find it difficult to make sense of these numbers, the **expected frequency tree** (Figure 1.1) may help to clarify them:

Figure 1.1 Expected frequency tree for doping example, showing that a test that is claimed to be '95% accurate' can still generate more false-positives than true detections: out of 68 positive tests, we would expect only 19 are truly doping

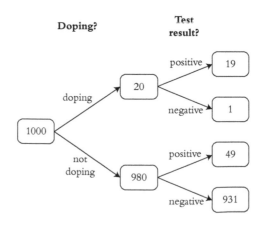

1.6 Teaching probability

Probability is vitally important but poorly understood, and therefore good teaching is essential. It forms part of most secondary school mathematics curricula and is also a component of many science curricula. However, the problems associated with the teaching and learning of probability are well documented. Over 20 years ago, Garfield & Ahlgren [9] noted a number of reasons for this, including difficulty with proportional reasoning and interpreting verbal statements of problems; conflicts between the analysis of probability in the mathematics lesson and experience in real life; and premature exposure to highly abstract, formalised presentations in mathematics lessons. Teacher knowledge may also be an issue, since not all teachers will have studied probability during their own education.

We might also add the continued focus on permutations and combinations, a topic that we (and we believe we are not alone) find intensely tedious. These are often seen as part of a probability curriculum and yet have nothing do with probability itself, being simply tools for counting possible outcomes.

We have sympathy with the struggle for comprehension by both teachers and students. When confronted by a school-level algebra question, students should know the steps required to work through to the answer. Probability questions are different: they always require careful thought, and the precise wording is crucial, as it is easy for it to be ambiguous. We personally try to check answers using at least two different solution methods.

At worst, probability can be taught purely in terms of abstract ideas, for example in this question from a (nameless) examination board.

> Consider three events A, B and C. A and B are independent, B and C are independent and A and C are mutually exclusive. Their probabilities are P(A) = 0.3, P(B) = 0.4 and P(C) = 0.2. Calculate the probability of the following events occurring: (i) Both A and C occur. (ii) Both B and C occur. (iii) At least one of A or B occur.[3]

[3] *For the solution of this horrible question, see Chapter 17 on independent events.*

Here, lack of any connection with the real world means that mistakes are difficult to spot, as it is impossible to apply common-sense ideas of magnitude. Fortunately, most examination boards manage somewhat more engaging questions.

1.7 Experimentation and modelling

Our approach is to teach probability through experimentation, and to use mathematical models to solve contextualised problems. We do not make a big issue about whether probabilities are 'known' or 'unknown'. In real life no probabilities are ever known, they are only assumed with more or less justification, and assessing probabilities from data through statistical inference is not the concern of this book.

We regard experimentation as vital to understanding the role of chance and unpredictability. Ideally students should carry out experiments themselves using randomising devices. Our preference is for spinners, either where probabilities are obvious or where they are deliberately concealed – these better reflect real life where nicely balanced situations are rare. We prefer to avoid dice – the choice of outcomes is too restrictive, and numbers may have emotional connotations, quite apart from the practical problems of throwing them in a class.

Having acquainted themselves with spinners as randomising devices in their own right, students can start using these as a way to model real situations – spinners can easily be labelled with specific outcomes, and students then simply count the number of times outcomes occur in a given number of trials. The idea is to use whole numbers initially, and bring in proportions, fractions and probability rules later. We also exploit the strong motivating role of playing competitive non-gambling games of chance, encouraging engagement by clearly making some outcomes more desirable than others.

1.8 What's in the book

Part 1 continues by introducing our approach to teaching probability in Chapter 2, together with our probability curriculum in Chapter 3. We do not follow any specific syllabus, although we have been influenced by the revised GCSE Mathematics (9–1) for England and Wales (2015), but present what we feel is a logical way to develop students' conceptual understanding over a period of four or five years. Our curriculum is sub-divided into three levels, corresponding to the first year or two of secondary teaching, the middle year or two, and then the final year or two.

Part 2 presents a series of detailed classroom activities. The activities in Part 2 can be tackled at more than one level, and detailed notes are provided for this. We consider it advantageous for students to study a good scenario in depth, revisiting it to discover how it can be interpreted in a more advanced way.

Part 3 works through an extensive series of sample assessment questions, with multiple solution methods wherever possible. Inspiration for the style and content of the questions is primarily from the sample assessment material provided by examination boards for the revised GCSE Mathematics (9–1).

Part 4 presents a range of supplementary, extension projects in probability, including both classroom activities and mathematical explanations. We include the 'classics', such as matching birthdays, lotteries, patterns of randomness, and Monty Hall, but we also feature various games, explorations of misconceptions about probability, the idea of 'fairness', psychological attitudes to risk, and in particular the misleading way that risks are often communicated in the media.

1.9 References

1 Galesic M, Garcia-Retamero R. *Statistical Numeracy for Health: A Cross-cultural Comparison With Probabilistic National Samples.* Archives of Internal Medicine. 2010 Mar 8;170(5):462–8.

2 Easton M. *What happened when MPs took a maths exam* [Internet]. BBC News. [cited 2015 Nov 11]. Available from: http://www.bbc.com/news/uk-19801666

3 Gigerenzer G, Hoffrage U. *How to improve Bayesian reasoning without instruction: Frequency formats.* Psychological Review. 1995;102(4): 684–704.

4 Gigerenzer G, Edwards A. *Simple tools for understanding risks: from innumeracy to insight.* BMJ. 2003 Sep 27;327(7417):741–4.

5 Gigerenzer G, Gaissmaier W, Kurz-Milcke E, Schwartz LM, Woloshin S. *Helping Doctors and Patients Make Sense of Health Statistics.* Psychological Science in the Public Interest. 2007 Nov;8(2):53–96.

6 Martignon L, Kurz-Milcke E. *Educating children in stochastic modeling: Games with stochastic urns and colored tinker-cubes.* Available from: https://www.unige.ch/math/EnsMath/Rome2008/WG1/Papers/MARTKU.pdf

7 Reyna VF, Brainerd CJ. *Numeracy, ratio bias, and denominator neglect in judgments of risk and probability.* Learning and Individual Differences. 2008;18(1):89–107.

8 NHS Breast Screening Programme. *Breast screening: helping women decide* [Internet]. [cited 2015 Nov 11]. Available from: https://www.gov.uk/government/publications/breast-screening-helping-women-decide

9 Garfield J, Ahlgren A. *Difficulties in Learning Basic Concepts in Probability and Statistics: Implications for Research.* Journal for Research in Mathematics Education. 1988;19(1):44–63.

Probability in the classroom

We want students to develop their understanding of probability, as well as technical competence. The following steps are those we have identified as providing this understanding – these steps are first listed below and then described in more detail.

1 Start with a problem, expressed as an appropriately simplified mathematical model or game.

2 Use a spinner as a tool for investigating the model, where each spin generates an observed result.

3 Do experiments in small groups to promote discussion, recording the results of the spins either on paper or using physical objects, such as multi-link cubes.

4 Tally the data, then record it on a frequency tree.

5 Discuss the narratives represented by sets of branches on the tree, emphasising that these are mutually exclusive, and that together they encompass all possible outcomes in the experiment. The list of outcomes is the sample space for the experiment.

6 Ask questions about the proportion of times specific events occur, and whether and why any results are surprising.

7 Average the data from all groups, observing that this 'smooths' the data, so that trends in the data can be seen more clearly.

8 If possible, conduct large numbers of trials using a computer animation, helping students to understand that the more experimental results they have, the nearer the data approaches the results they would expect if they could conduct an infinite number of trials.

9 Construct the expected frequency tree, discussing the proportion of times you expect each outcome/event to occur. Compare with the experimental data – the class average may well be very close to the expected results.

10 Compare representations of data in:

    a   frequency trees

    b   contingency (2-way) tables

    c   Venn diagrams.

11 Decide what fraction of times you expect each outcome on the spinner, and use these fractions as probabilities on tree branches to arrive finally at the probability tree.

Note that probability is the *final* step in our teaching method. The preceding steps provide scaffolding for students to build up their experience of what happens experimentally, and provide many opportunities to discuss what different representations tell us, before they are faced with the abstract probability tree. Note too that we do not ask students to predict what will happen before they have had a chance to see what happens experimentally. We do not want them to form unfounded opinions which then blind them to the evidence.

2.1 Starting with an experiment

In this chapter we are going to use a blue/yellow spinner where the blue area is greater than the yellow area. (All the spinners in this book are available online in colour and labelled grey-scale versions.)

Image 2.1 The spinner

Pull out one end of a paperclip, then pin the loop of the paperclip to the centre of the spinner with a pen or sharp pencil. Flick the extended end of the paperclip. It should spin rapidly, finally coming to rest in either a blue region or a yellow region (in Image 2.1 the paperclip has finished in a yellow region). If the result is unclear, simply spin again.

We are going to use this spinner as a tool to explore a very simple mathematical model: what is the weather going to do? However, before we launch into an experiment to collect data to investigate this model, we need to clarify key terms.

Suppose we define blue on the spinner to mean a rainy day, and yellow to mean a sunny day. These are the only two possible **outcomes** of a single spin. The list of the outcomes (blue = rainy, R; yellow = sunny, S) is the **sample space**.

In a mathematical model, we start with a very simple model, and then build up complexity. Suppose we decide that a better model would be to differentiate between morning and afternoon. We now have four outcomes: RR (rainy in the morning and in the afternoon), RS (rainy in the morning, sunny in the afternoon), SR (sunny in the morning, rainy in the afternoon) and SS (sunny in the morning and in the afternoon). The sample space for this model is the set of outcomes RR, RS, SR and SS.

To use the spinner to collect experimental data, we need one further piece of information – how many **trials** will be needed to collect the required data? For our model, a trial consists of spinning the paperclip on the spinner twice, once for the morning and once for the afternoon. For reasons which will become clear in due course, we suggest 25 days.

Suppose we want to know on how many occasions in the 25 days it rains at some point in the day. This means finding the total frequency for RR + RS + SR. That gives us the frequency of an **event**, which is a specific set of outcomes and a subset of the sample space. Other events that we might count include the incidence of totally sunny days (SS), or the incidence of days in which the weather in the morning is different from that in the afternoon (RS + SR).

We have introduced the language needed to carry out the experiment: what a trial is (two spins of the spinner) and how many trials we will do (25). We have introduced the language needed to report the data: the result of one trial is an outcome (e.g. rain in the morning and rain in the afternoon, RR), and the list of all possible outcomes of each trial is the sample space (RR, RS, SR, SS). It is important for students to be aware that the sample space includes all possible outcomes, even if one or more outcomes do not occur in a particular experiment. If we want to say something about a set of outcomes making up a subset of the sample space (which includes individual outcomes and the whole sample space), then that is an event.

Although the spinner and weather model are quite simple, we have not exhausted their potential yet. Going back to the first model, where we assigned only one type of weather to each day, data for 25 days will look something like this:

Sequence 1:   R R S S   R S R S S   S R R S S   R S R S S
                 S R R S S

Sequence 2:   R R R R R   R R S R S   R R R R R   R R S S R
                 R S R S R

One of these sequences was generated by a human being, the other by a computer. Can you tell which is which?

Generally when we try to generate random sequences, we try too hard to avoid obvious patterns, and we instinctively want to balance the outcomes. Sequence 1 is human generated, while Sequence 2 was generated using the RANDOM function in Excel. Sequence 2 has considerably longer runs (of R in this case), with two runs of seven Rs compared to the runs of three Ss in Sequence 1.

It is important to give students experience of random sequences, and not to invent 'random' results. Common misconceptions are that if an outcome has not happened recently, say, a 6 on a die, then one is due, or alternatively that the die is biased – both these can be tackled by looking at random sequences. Another common misconception is to consider a result highly significant, like the two runs R of length 7, or getting a Head six times in a row, when in reality that is no more remarkable than any other specific sequence. It is we who see more significance in HHHHHH than in HTTHTT – both are equally likely to occur when a coin is spun six times. See *What does 'random' look like?* in Chapter 21 for classroom activities on the patterns of randomness.

Can we say anything about how likely a sunny day or a rainy day is from the spinner or the data? Not really. The spinner has a greater blue area than yellow, but it is not clear what proportion there is of either. Sequence 1 is useless, because it is not a random sequence at all, and Sequence 2 does not provide enough information.

Table 2.1 shows a set of simulated results (generated in Excel) for a class of 30 students working in pairs:

Group	1	2	3	4	5	6	7	8	9	10	11	12	13	14	15	Total	Average
R	18	11	12	15	15	18	19	17	16	14	10	15	17	12	15	224	14.9
S	7	14	13	10	10	7	6	8	9	11	15	10	8	13	10	151	10.1

Table 2.1 Testing the spinner

This suggests that R is likely to occur on $\frac{15}{25}$ of the occasions, and S on $\frac{10}{25}$, which provides one estimate for the proportion of blue and yellow areas on the spinner.

2.2 Recording experimental data

While random sequences are interesting in their own right, data recorded as a sequence does not facilitate further analysis. Table 2.2 shows a data set generated by the blue/yellow spinner and recorded as representing the weather for the morning and afternoon over 25 days.

1		2		3		4		5	
	SR		RS		SR		RS		SR
6		7		8		9		10	
	RR		SS		RS		SR		SR
11		12		13		14		15	
	RS		RS		RS		RR		SR
16		17		18		19		20	
	RR		RR		RR		SR		RR
21		22		23		24		25	
	RR		RS		SR		RR		SS

Table 2.2 Data for the weather-modelling experiment

Presented like this, as a sequence of outcomes, the data is quite difficult to interpret, so the next step is to tally the data (Table 2.3).

Outcome	Tally	Total
SS	//	2
SR	///// ///	8
RS	///// //	7
RR	///// ///	8
		25

Table 2.3 Tally for weather experiment

This information is then recorded on a **frequency tree,** which provides a structure to enable students to get inside their data and to start analysing what it is telling them (see Figure 2.1).

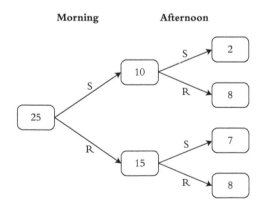

Figure 2.1 Data recorded on a frequency tree

This may resemble a probability tree, but it is not. You, as a teacher, will already be familiar with probability trees, but you need to put them out of your mind for the time being! Remember – if your students are coming to this new, it is simply a way to help them record what happens whenever there is a random outcome, i.e. whenever they use the spinner. A frequency tree is a means of displaying data, so the numbers are whole numbers. The arrows on the branches emphasise that there is a progression from left to right, in this case from morning to afternoon.

The tally table (Table 2.3) in its present form gives students the numbers that go in the right-hand boxes – the frequencies of the final outcomes – but it does not make it easy to see what should go in the intermediate boxes – the results for the morning. Discussion of what the numbers in those boxes represent – the frequency of S or R for the morning, yellow or blue on the first spin, regardless of what followed – will help students not only to complete this tree, but also to understand the tree structure. Using a frequency tree to record data is a first step to ensuring that the later introduction of probability trees goes smoothly. At this introductory stage do not mention probabilities at all, but simply use the tree structure to record data in whole numbers.

Looking at the progression of events on the tree, we can understand it as four mutually exclusive narratives. These should be unpacked with the class. This dialogue is based on a real exchange between a teacher and her class (a similar exchange is reported in the higher-level analysis of Chapter 5).

Teacher: Look at the top branches. What's the story here? (Image 2.2.)

Student: It's sunny in the morning, and sunny in the afternoon.

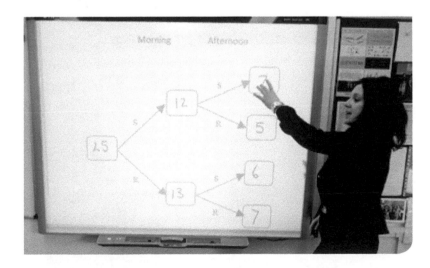

Image 2.2 What do the top branches represent?

T:	And how about these two branches? (Image 2.3.)
S:	Sunny in the morning, rainy in the afternoon.
T:	And these two branches? (Image 2.4.)
S:	Rainy in the morning, sunny in the afternoon.
T:	And these? (Image 2.5.)
S:	Rainy in the morning, rainy in the afternoon. Better take an umbrella!
T:	So we've got four possible outcomes. What are they?
S:	Sunny all day, sunny then rainy, rainy then sunny, and rainy all day.

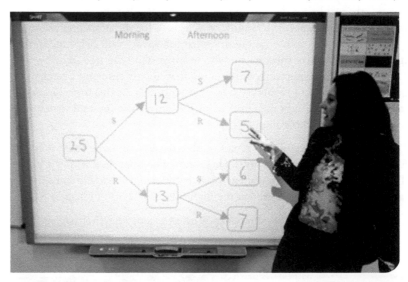

Image 2.3 What do the next branches represent?

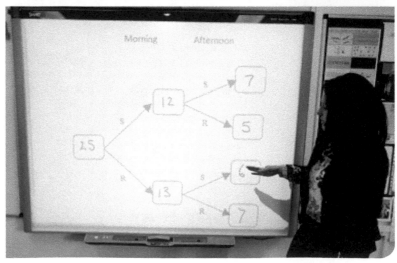

Image 2.4 And these two branches?

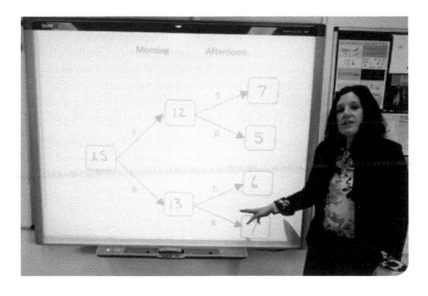

Image 2.5 And these?

T: Could anything else happen if we stick with this model?

S: No.

A dialogue like this helps students to understand both the sequence of events and the tree structure, so that they can be quite sure that all possible outcomes are included. Students should always explicitly identify the sample space: the complete list of mutually exclusive outcomes.

2.3 From empirical results to expected results

So far, we have worked entirely with experimental data. We have deliberately used a mathematical model for this, as the difference between, say, a rainy morning and sunny afternoon, and a sunny morning and rainy afternoon, is so much more obvious than between a Head and a Tail, and a Tail and a Head, on a coin. The spinner chosen also means we do not start from any assumptions about outcomes being equally likely – there is clearly more blue than yellow.

The next step is to look at the average results of an experiment across a whole class. Students should be encouraged to compare the most extreme sets of results in the room with the mean results obtained by averaging every group's results and rounding answers to one decimal place. Through discussion, students should be able to convince themselves firstly that the average of a number of groups provides a better estimate of what would happen if they could do a very large number of trials, and secondly that this average can also be used to estimate what proportion of the spinner is blue and what proportion is yellow.

The spinner is actually 60% blue and 40% yellow. What does this tell us? For the simple model, where we assume that the weather will stay the same all day, it means that if we could do a large number of trials, then we would expect blue = rainy day 60% of the time, and yellow = sunny day 40% of the time. The more trials we do, the closer we expect to get to this. The **expected result** for the weather model with a 25-day period, therefore, would be 60%, or 15, rainy days and 40%, or 10, sunny days.

To calculate what we expect for the model where the weather can differ between morning and afternoon, we need to construct an **expected frequency tree** (Figure 2.2).

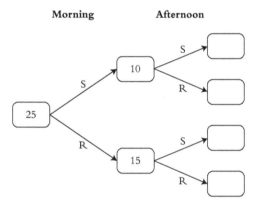

Figure 2.2 Constructing an expected frequency tree

In Figure 2.2, the boxes for the expected results for the morning have been completed – this should be straightforward. To fill in the afternoon boxes, it is helpful to think again about the narrative structure of the tree. What does that top set of branches mean? It means a day when it is sunny in the morning, and then sunny again in the afternoon. We have 10 days when we expect sun in the morning, so the question is – for how many of those 10 days do we also expect sun in the afternoon? The spinner is the same, so we expect 40% of the 10 sunny mornings to also be sunny in the afternoon, i.e. 4 SS days, and we expect 60% of the 10 sunny mornings to become rainy in the afternoon, i.e. 6 SR days. (Sunny, which is less likely than rainy, is deliberately placed at the top, so that the first calculation we have to do here is the easiest one. The 25-day period was chosen to ensure that these calculations do not involve fractions or decimals at this stage.) Similar calculations for the 15 days when we expect rain in the morning give 6 RS days and 9 RR days (there are various ways of facilitating this calculation for students so that it is a mental arithmetic exercise rather than a calculator-aided one).

Figure 2.3 shows the finished expected frequency tree.

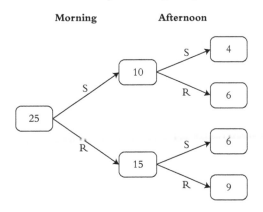

Figure 2.3 A completed expected frequency tree

What proportion of days do we expect to be sunny in both the morning and the afternoon? 4 out of 25, or $\frac{4}{25}$. We are on the way to understanding that the **probability** that a day is sunny throughout (in this model) is $\frac{4}{25}$.

It is important that students realise the difference between recording experimental data on a frequency tree and constructing the expected frequency tree, which is a means of making predictions. This is essentially the difference between observed relative frequency and theoretical probability.

2.4 Alternative representations

Once students are comfortable with representing data on a frequency tree, then 2-way tables and Venn diagrams should also be used. All three are ways of recording exactly the same data, but the different representations draw attention to different aspects of the data.

The tree structure emphasises the sequence of events, the number of different narratives and the complete set of mutually exclusive outcomes which make up the sample space. The right-hand boxes give the frequency of each outcome. If we want to know the frequency of an event such as 'rainy morning', then we have to look at the intermediate box labelled R, at the end of the appropriate 'morning' branch. If we want to know the frequency of 'sunny afternoon', we have to add the frequencies in the two right-hand boxes labelled SS and RS.

In the 2-way tables (Table 2.4 and Table 2.5), the four outcomes appear in the central, shaded cells. The Total row (bottom), shows the number of days in which the morning was either sunny (S-) or rainy (R-). The Total column (far right) shows the number of afternoons that were sunny (-S) or rainy (-R). This way of representing the data emphasises the events corresponding to the row and column totals, as well as the four mutually exclusive outcomes. Students should identify corresponding events on the tree and on the 2-way table – both contain exactly the same information, but displayed quite differently. The 2-way table loses the time-ordered narrative structure, but makes it easier to see the different combinations of events.

		Morning		Total
		S	R	
Afternoon	S	SS	RS	-S
	R	SR	RR	-R
Total		S-	R-	

Table 2.4 2-way table showing structure of weather model

		Morning		Total
		S	R	
Afternoon	S	2	7	9
	R	8	8	16
Total		10	15	25

Table 2.5 2-way table showing empirical data for weather experiment

Setting up a Venn diagram is not trivial. How should the circles be labelled? We label the branches of the frequency tree according to how we conducted our experiment – so branches for sunny and rainy in the morning, followed by branches for the choice between them for the afternoon. We label the rows and columns of the 2-way table similarly, with the possible choices for the morning in the columns (they could equally well go on the rows), and the possible choices for the afternoon in the rows (or columns). It does not take long to realise, however, that if we label the two circles of the Venn diagram sunny and rainy, we are in trouble – what would the intersection mean, for a start?

It is crucial that students realise that the four regions on the Venn diagram correspond to the four mutually exclusive outcomes – SS, SR, RS and RR – and that it is easiest to start by labelling the regions with the outcomes. Labelling the circles then follows, by seeing what is common to the two labels each circle contains. The two Venn diagrams in Figure 2.4 are both correct and equivalent ways to represent the data.

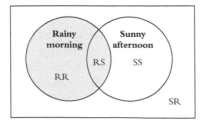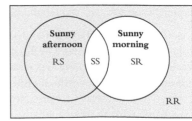

Figure 2.4 Labelling a Venn diagram

20

If we want, say, to find the frequency of the event 'rainy morning', we add the frequencies in the appropriate two regions, shaded on the diagrams in Figure 2.4, so RR + RS = 8 + 7 = 15 (Figure 2.5).

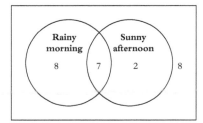 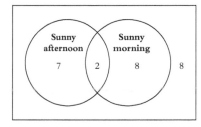

Figure 2.5 Venn diagram showing empirical data for weather experiment

It is well worth spending time *comparing* the frequency tree diagram, the 2-way table and the Venn diagram in terms of the outcomes, and looking at events formed from sets of outcomes, so that students realise that these representations do in fact display the same data, and also understand why we add frequencies for events consisting of more than one outcome.

2.5 From expected results to theoretical probabilities

We do not start with probability trees, because they present students with a whole range of new concepts and techniques simultaneously, and many problems can be solved without them, using just frequencies (see Part 3). Starting with the frequency tree and empirical data familiarises students with the tree structure. We then introduce the expected frequency tree, focusing on proportions to answer questions. In the weather experiment, we used a 25-day period, because the calculations of proportions for the morning and then for the afternoon could be solved using only whole numbers. However, the choice of time period is arbitrary: in any experiment we need to specify how many trials there will be, but eventually we need to abstract the calculations from the experimental context so that we are not restricted to that number of trials.

On a frequency tree, the sum of the frequencies for the outcomes must be the same as the number of trials. The sum of the probabilities of the outcomes is 1, so in moving from the expected frequency tree to a probability tree, we have implicitly generalised the experiment. Instead of working with a number of trials, we agree that, whatever period might be chosen, it represents 100% of the trials. However, working with percentages on a probability tree should be avoided with students, because it is so easy to forget that a percentage is actually a fraction, not a whole number.

When moving from an expected frequency tree to a probability tree, we remove the boxes for results, and place the labels directly at the ends of the splits in the branches (see Figure 2.6). We also remove the arrows we

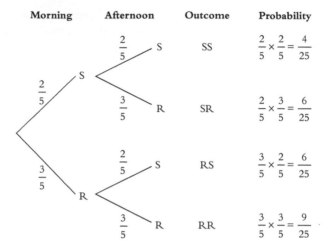

Morning	Afternoon	Outcome	Probability

$$\frac{2}{5} \times \frac{2}{5} = \frac{4}{25}$$ for SS

$$\frac{2}{5} \times \frac{3}{5} = \frac{6}{25}$$ for SR

$$\frac{3}{5} \times \frac{2}{5} = \frac{6}{25}$$ for RS

$$\frac{3}{5} \times \frac{3}{5} = \frac{9}{25}$$ for RR

Figure 2.6 Probability tree showing probabilities of outcomes derived from the expected frequency tree in Figure 2.3

put on the branches of the frequency tree, helping to emphasise that this is a different tree. The probabilities on the individual splits in the branches represent the number of times each event is expected as a fraction of the total number of trials represented by that split. The probability at the end of each entire branch represents the overall number of trials expected to result in that outcome as a fraction of the total number of trials.

Multiplying along the branches follows from the calculations we did on the expected frequency tree. The proportion of sunny mornings was 40%, and the proportion of sunny mornings followed by sunny afternoons was 40% of 40%, which is the same as multiplying $\frac{2}{5}$ by $\frac{2}{5}$. If we want to know the probability of the event 'a sunny afternoon', then we need to add the probabilities for the relevant outcomes (i.e. for SS and RS) just as we added frequencies on the expected frequency tree.

2.6 The theory of probability

It is rather remarkable that the approach described above essentially covers not only the theory of probability required in the curriculum, but all the theory of probability that there is to know!

The basic rules of probability, given in mathematical notation, can be expressed as follows.

Definition:

> *The probability of an event* A, *denoted* P(A), *is a number between 0 and 1, with* P(A) = 0 *corresponding to* A *being impossible, and* P(A) = 1 *to* A *being certain.*

Complement rule:

> P(*not* A) = 1 − P(A)

Addition rule:

for mutually exclusive events A, B: $P(A \text{ or } B) = P(A) + P(B)$

for non-mutually-exclusive events A, B: $P(A \text{ or } B) = P(A) + P(B)$
$$- P(A \text{ and } B)$$

Multiplication rule:

for independent events A, B: $P(A \text{ and } B) = P(A) \times P(B)$

for dependent events A, B: $P(A \text{ and } B) = P(A|B) \times P(B),$
where $P(A|B)$ *is the conditional probability of*
A *given* B.

A crucial idea is that all these rules follow intuitively when our raw information is provided as a set of data, and we are imagining what we might see if we pick a single instance 'at random' from this set.

Consider, for example, the frequency tree in Figure 2.1, which describes what happened in 25 trials of an experiment. Suppose we were to ask: if I take one of these trials at random, what is the probability that it will be 'SS'? Simple enumeration gives an answer of $\frac{2}{25}$. Similarly, the probability that it is 'not SS' equals $\frac{23}{25}$, which is 1 minus the probability it is 'SS', and so we have derived the complement rule. The other rules follow from simple set theory.

But, rather than work with these experimental probabilities derived from an *observed* frequency tree, we can derive the rules for theoretical probabilities by working with *expected* frequency trees. We have so far taken an informal approach to the idea of an expected frequency tree, relying on the intuition that if, for example, we have a spinner with 40% yellow and 60% blue, then if we spin it 10 times we 'expect' 4 yellows and 6 blues. More formally (but not introduced to students until much later):

*If we have n events, each with probability p of occurring, then the **expected** total number of occurrences is n × p (this holds even if the events are not independent).*

Having produced an expected frequency tree, we think of it as a population of future outcomes in the proportions in which we expect them to happen. The probabilities associated with a single future outcome follow from thinking of it as a random observation from this population.

All this sounds very difficult and technical, but once students are familiar with expected frequency trees and comfortable working with them, the rest of the theory of probability that they require should follow in a fairly straightforward manner.

In an expected frequency tree, the sum of the results for all the outcomes must equal the number of trials, and so, for example, in Figure 2.3 the sum of the expected frequencies of SS, SR, RS and RR must add to 25; when we transform to a probability tree, the sum of the probabilities for all the outcomes must equal 1. In particular, if P(event) = p, then P(not event) = $1 - p$, and so we have the complement rule.

Similarly, when considering a composite event such as 'rainy morning or afternoon', its frequency is clearly the sum of the frequencies of its constituent events SR, RS and RR, or $6 + 6 + 9 = 21$ in Figure 2.3. When transforming to the probability tree, we have P(rainy morning or afternoon) = P(SR) + P(RS) + P(RR) = $\frac{21}{25}$, the addition rule for probabilities of mutually exclusive events. We can also see this as an example of non-mutually-exclusive events, since 'rainy morning or afternoon' comprises the set 'rainy morning R–' plus the set 'rainy afternoon –R', minus the set 'rainy morning and afternoon RR', and so P(rainy morning or afternoon) = P(R–) + P(–R) – P(RR) = $\frac{15 + 15 - 9}{25} = \frac{21}{25}$.

The multiplication rule for composite events generalises the process of taking a proportion of a proportion: for our blue and yellow spinner, we expect 60% or $\frac{3}{5}$ of spins to give blue, so for the weather model we expect RR $\frac{3}{5}$ of $\frac{3}{5}$ of the time.

For our weather experiment, the afternoon event is **independent** of that in the morning, since the probability of rain in the afternoon does not depend on what happened in the morning. Independence is not just a feature of mathematical models using spinners (or dice or coins or playing cards): if I bought a lottery ticket yesterday and did not win a prize, that will not affect the probability that I win a prize on the ticket I buy today – there is no sense in which I am 'due' a prize because I lost yesterday, or any sense in which I am less likely to win today because I lost yesterday and am clearly a loser where the lottery is concerned!

However, many events in real life are not independent. Our simple weather model could be improved if we consider that the weather in the afternoon might depend on the weather in the morning. If it rains in the morning, then perhaps the chance of rain in the afternoon should be set higher than 60%. The experiment could be adapted by using a different spinner for the afternoon. The activity *The dog ate my homework!* (Chapter 6) has different probabilities in the second stage of the experiment, depending on what happens in the first stage. It becomes clear that the rule for multiplying the probabilities on a branch in order to find the overall probability of an outcome holds whether the events are independent or not. *Choosing representatives* (Chapter 7) provides a context for exploring independent and dependent probabilities further.

Conditional probability is generally considered to be a difficult topic, to be covered right at the end of a matriculation or GCSE course, or left for a more advanced course such as GCE A Level. This is a pity, since such questions are very accessible using the frequency approach advocated in this book. *The dog ate my homework!* (Chapter 6) provides students with opportunities to discuss conditional probability in an informal way, and to

answer questions without needing to resort to a formula. This can easily
be extended to more sophisticated conditional probability questions
of the type illustrated by the 'doping' example in Chapter 1, which are
essentially applications of **Bayes theorem**. These classically difficult
problems become reasonably straightforward using an expected frequency
approach.

Probability notation can also be used to emphasise the similarities and
differences between the three commonly used representations: the
probability tree, the 2-way table and the Venn diagram. Suppose we have
two events, Event 1 and Event 2, say first coin toss and second coin toss,
and for each the outcome could be A or B, Heads or Tails. (See Figure
2.7, Table 2.6 and Figure 2.8.)

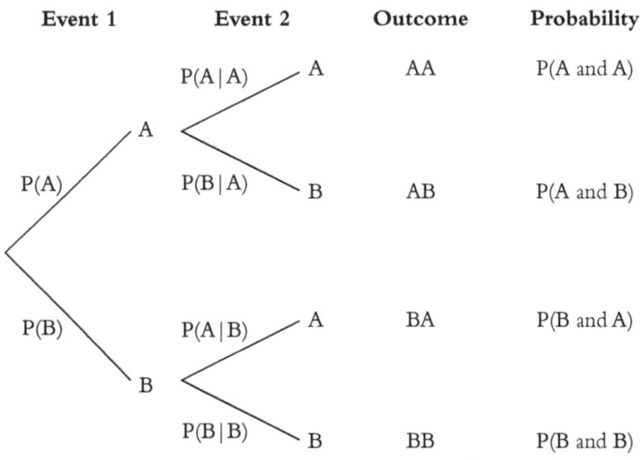

Event 1	Event 2	Outcome	Probability
	P(A|A) → A	AA	P(A and A)
A	P(B|A) → B	AB	P(A and B)
P(A)			
P(B)	P(A|B) → A	BA	P(B and A)
B	P(B|B) → B	BB	P(B and B)

Figure 2.7 Probability tree for two events

		Event 1		A or B
		A	B	
Event 2	A	P(A and A)	P(B and A)	P(Event 2 = A)
	B	P(A and B)	P(B and B)	P(Event 2 = B)
A or B		P(Event 1 = A)	P(Event 1 = B)	1

Table 2.6 2-way table for two events

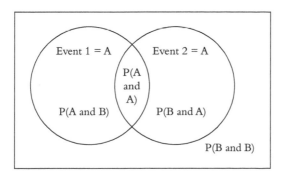

Figure 2.8 Venn Diagram for two events

25

Note that Event 1 = A applies to the whole of the left-hand circle in Figure 2.8, and Event 2 = A applies to the whole of the right-hand circle. From this we can deduce what combination of events each region refers to. It is conventional to use set notation for the contents of the regions of a Venn diagram, where the intersection would be the set (Event 1 = A) ∩ (Event 2 = A). We do not use this notation in this book, as it is not necessary for the theory of probability at this level.

> The running theme is to realise the importance of being absolutely clear about what a probability really refers to. This clarity is achieved by expressing a probability as a fraction, with the denominator corresponding to the size of the sample or population in which events of interest may occur, and the numerator being the expected number of occurrences of that event.

We hope that the material presented here will encourage that clarity.

Essentially we think of probability in terms of the more intuitive idea of expectation, rather than the standard process in which probability comes first.

The probability curriculum

In this book we cover the requirements of probability curricula for school students up to matriculation (GCSE Higher Tier) level. The activities are discussed at three levels: introductory, intermediate and higher.

Introductory-level material is intended to be suitable for the first year or two of secondary school for most students. Intermediate-level material builds on this, taking students up to Grade/Year 9 or 10. Higher-level material is appropriate for students preparing for matriculation or GCSE Higher Tier; GCSE Foundation Tier may require some of this material, but not all of it will be needed.

The three sections in the remainder of this chapter outline the content of each of these three levels. There is a table for each level, with examples intended to clarify the content, and additional notes where needed. Most of the examples use one or more of the spinners shown in Figure 3.1.

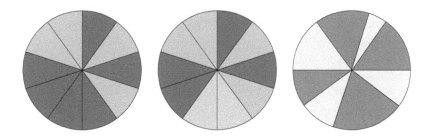

Figure 3.1 Spinners

The left-hand spinner has 10 equal sectors, 6 blue and 4 green. The middle spinner also has 10 equal sectors, 4 blue, 4 green and 2 orange. The right-hand spinner has an indeterminate number of equal sectors, of which more are red than are yellow. These spinners are intended to be used with a paperclip as explained at the beginning of Chapter 2.

The examples in this chapter are provided simply as illustrations, and not as part of a scheme of work. Part 2 of this book contains activities intended for use in a work scheme, including initial planning and use in the classroom, in addition to discussion and analysis at the appropriate levels.

3.1 Introductory level

Curriculum content	Example	Notes
Verbal probability scale: impossible, evens (50-50), certain, more or less likely	With the blue/green spinner, is a paperclip more likely to land on a blue sector or a green one? What about the blue/green/orange spinner?	It is important not to equate words such as 'likely' with a specific numerical value. It is better to use them as comparative terms.
Simple numerical scale: impossible = 0, certain = 1, evens = $\frac{1}{2}$ or 0.5	On a number line from 0 to 1, show the probability of getting green on the blue/green spinner.	
Simple events: what is the chance of a favourable outcome from a small number of equally likely outcomes?	If a paperclip is spun around the blue/green spinner, what is the chance that it will land on a green sector? If a paperclip is spun around the blue/green/orange spinner, what is the chance that it will land on an orange sector?	The number of equally likely outcomes is 10 in both cases, not 2 or 3. The chance of a favourable outcome from all equally likely outcomes is the probability of that outcome. The informal term 'chance' is used at this introductory level.
Random sequences	The blue/green spinner gave these results: B G G G G What is the chance that it will be green on the next spin?	Understanding that a spinner (like a coin or a die) has no memory – blue is not due, nor is green more likely because there has been a short run of greens.
Frequency trees as a recording tool	Spin the blue/green spinner twice, recording the result of each spin. Repeat 20 times, then record your results on a frequency tree (Figure 3.2). **Figure 3.2** Example of frequency tree for the B/G spinner	Recording frequencies on a tree helps students to understand the tree structure before they need to use it for calculating probabilities. A frequency tree has frequencies in boxes at the ends of branches, with the outcome of each spin on the branch and arrows to emphasise the progression.

(continued)

Curriculum content	Example	Notes
Estimating probability from experiment (relative frequency)	Spin the yellow/red spinner a number of times. The proportion of each colour obtained provides an estimate for the probability that on any given spin, the paperclip will stop at that colour.	Students should be encouraged to discuss variability in results, and to agree that the greater the number of trials, the better the probability estimates are likely to be. If a class is divided into a number of groups, and groups' results are averaged, this is likely to give a better estimate of probabilities, without each group needing to do hundreds of trials. Computer simulations are also appropriate for this.
Complement – frequency and proportion	If the proportion of times that the paperclip stopped at a yellow section of the red/yellow spinner was 20 out of 50 ($\frac{20}{50}$), what was the proportion of times that it stopped at a red section?	This question could be asked in terms of frequency first.

Good activities for this level are *Fair Game?* (Chapter 4) and *Which team will win?* (Chapter 5).

3.2 Intermediate level

Once a class is comfortable with the introductory content, they can move on to the intermediate-level.

Curriculum content	Example	Notes
Mutually exclusive events – frequency and proportion	If the blue/green/orange spinner is spun 50 times, and blue and orange occurred 12 and 20 times respectively, how many times did the spinner stop on green? What proportion is that?	This extends the concept of complement to more than two categories.
Non-exclusive events – frequency and proportion	Spin the blue/green spinner and the blue/green/orange spinner the blue/green spinner and the blue/green/orange spinner 10 times, and count how often blue came up on *at least one* of the spinners.	Students should note that the frequency of 'at least one blue' equals the frequency of blue on spinner 1 plus the frequency on spinner 2, minus the frequency on both. They can also express this on a Venn diagram.

(continued)

Curriculum content	Example	Notes
Recording complex events on a frequency tree	Spin the blue/green spinner, and then the blue/green/orange spinner, 25 times in total. Record your results on a frequency tree (Figure 3.3). 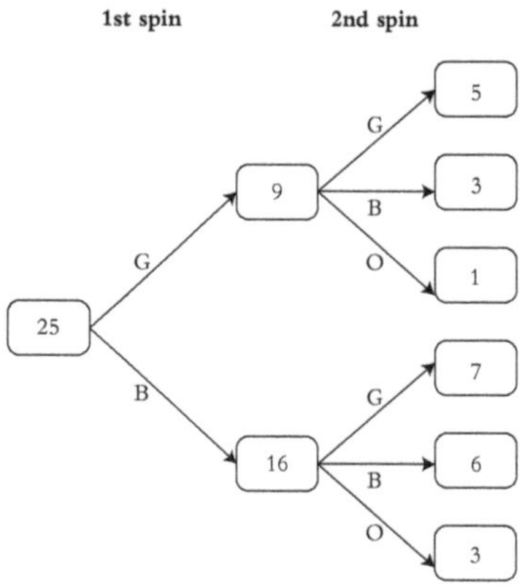 Figure 3.3 Example of frequency tree for B/G and B/G/O spinners	As well as completing the tree, students should list the outcomes, and recognise that these form the sample space for the experiment. They should also discuss what proportion of the 25 trials give, for example: GG; a blue and a green in either order; green first; green second; and so on. This will help them to get used to adding sets of branches for complex events.
Expected results recorded on a frequency tree and expressed as a proportion – two identical independent events	A trial in an experiment consists of spinning the blue/green spinner twice. If there are 50 trials, what proportion of these would we expect to result in GG? (See Figure 3.4.) 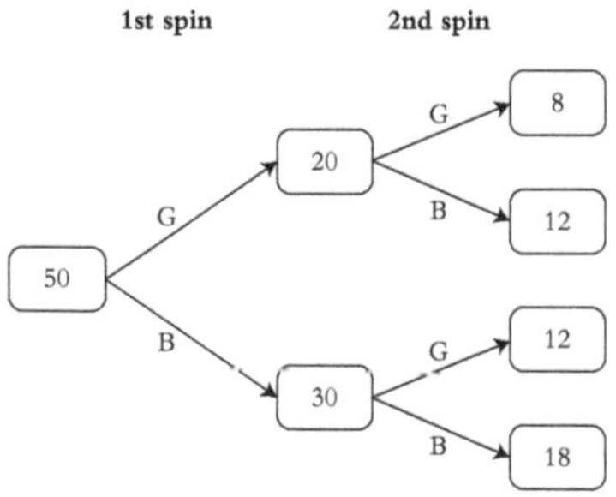 Figure 3.4 Expected frequency tree for B/G spinner	Working with an expected frequency tree is a vital intermediate stage between recording experimental results and moving on to a probability tree.

(continued)

Curriculum content	Example	Notes
Expected results recorded on a frequency tree and expressed as a proportion – two different independent events	A trial in an experiment consists of spinning the blue/green spinner once, then spinning the blue/green/orange spinner. If there are 50 trials, what proportion of these would we expect to result in GG? (See Figure 3.5.) 1st spin 2nd spin 50 → G → 20 → G → 8 20 → B → 8 20 → O → 4 50 → B → 30 → G → 12 30 → B → 12 30 → O → 6 **Figure 3.5** Expected frequency tree for B/G and B/G/O spinners	
Answering probability questions from expected frequency trees, where the probability is the number of favourable trials as a proportion of the total number of trials	What is the probability that if the blue/green spinner is spun, followed by the blue/green/orange spinner, that one will be blue and the other not blue?	Students can see from the tree in Figure 3.5 that the total frequency of this event is given by these branches: BG + BO + GB From this they can find the expected proportion for this event ($\frac{26}{50}$), which is equivalent to the probability of this event. This emphasises the need to add results for complex events (i.e. sets of branches) before introducing the multiplication rule.

(continued)

Curriculum content	Example	Notes
Sample space diagrams, which show all possible combinations of events, for enumerating outcomes with two or more attributes	What is the sample space if the blue/green and blue/green/orange spinners are both spun? (See Table 3.1.) Table 3.1 below **Table 3.1** Sample space for two spins of green/blue spinner	It is important to distinguish between a sample space diagram, which shows all the possible outcomes, and a 2-way table, which contains frequencies or probabilities.
2-way tables of expected frequencies	In the above experiment, how many outcomes are there? *6.* Are these equally likely? *No.* How much more likely is GG than GO? $\frac{8}{50}$ *compared to* $\frac{4}{50}$*, so twice as likely.* (See table 3.2.) Table 3.2 below **Table 3.2** Expected results of 50 trials of spins of green/blue and green/blue/orange spinners	
Explicit definition of expectation: expected number of events = probability of each event × number of opportunities for event to occur	For the blue/green spinner, the probability of 'green' is $\frac{4}{10}$, and so over 50 trials the expected number of 'greens' is $\frac{4}{10} \times 50 = 20$. In reverse, since we expect to get 20 'greens' in 50 trials, we know the probability of a 'green' is $\frac{4}{10}$.	This has been used intuitively to relate the expected number in a frequency tree to the underlying probability.

Table 3.1:

		1st spin	
		G	**B**
2nd spin	**G**	GG	BG
	B	GB	BB
	O	GO	BO

Table 3.2:

		1st spin		Total
		G	**B**	
2nd spin	**G**	8	12	20
	B	8	12	20
	O	4	6	10
Total		20	30	50

Good activities for this level are *Fair Game?* (Chapter 4), *Which team will win?* (Chapter 5) and *The dog ate my homework!* (Chapter 6).

3.3 Higher level

Classes preparing for public exams, such as matriculation or GCSE, will need some or all of this content, depending on the syllabus for the exam.

Curriculum content	Example	Notes
Two independent events (which may be the same or different) – probability tree	What is the probability that, if the blue/green spinner is spun twice, at least one is green? From the tree in Figure 3.6 below, P(at least one green) = P(GG) + P(GB) + P(BG) = 1 − P(BB) = 0.64 **Figure 3.6** Probability tree for the B/G spinner	The rule that we multiply along branches follows from the way in which expected frequencies are found for expected frequency trees.
Other representations	The above information could equally be given in a 2-way table or a Venn diagram.	These show the same information as the probability tree, but the different representations emphasise different aspects of the calculations.

The final topics at this level are better modelled using counters in opaque bags or boxes, to ensure that a counter drawn from the bag or box cannot be seen and so is a random choice.

Curriculum content	Example	Notes
Dependent events	An opaque bag contains 6 blue and 4 green counters. Two counters are taken out and not replaced. What is the probability that both are green? (See Figure 3.7.)	This is a standard probability question at this level, and is best tackled using a probability tree. This clearly

(continued)

Curriculum content	Example	Notes
	Figure 3.7 Multiplying probabilities along branches for dependent events The probability of two greens is $\frac{12}{90} = \frac{2}{15}$.	displays that the probabilities are multiplied along the branches, just as for independent events.
Conditional probability	Opaque Bag A contains 3 blue and 1 green counters. Opaque Bag B contains 1 blue and 3 green counters. I pick a bag at random, and take out a counter. If the counter is blue, what is the probability I have chosen Bag A? Figure 3.8 Using frequencies to solve a conditional probability problem From an expected frequency tree based on 8 repetitions of the experiment (as in Figure 3.8), it is clear that we expect the counter to be blue on 4 occasions, of which 3 are from Bag A. So the probability of having chosen Bag A in a single trial, given the counter is blue, is $\frac{3}{4}$.	Conditional probability questions are much more straightforward when approached using expected frequencies.

All the activities in Part 2 can be profitably used at this level.

Introduction to the resources

These activities provide resources for teaching probability in the way that we outline in Part 1. You could dive straight in, without reading Part 1 first, but these chapters will make a lot more sense, and be much more successful in the classroom, in the light of Part 1. How they are delivered is likely to make the difference between a successful lesson and one which is less so.

These activities have been used in the classroom, where we observed students learning through doing and gaining conceptual understanding without needing to have everything spelled out for them first. New exam regulations mean that students will need to be able to answer questions which they have not seen before, and where they need to apply their mathematical understanding, rather than relying on rote learning. Working together in groups, and having to articulate their thoughts to each other, helps students to develop independent learning skills, and the emphasis on practical activity and visual representations provides support for different learning styles.

The spinners used in these resources are available on the website (www.teachingprobability.org) as images, in colour and labelled grey-scale versions, for use with your own worksheets. They originated as pie charts in Excel, so there is also a spreadsheet with spinners which you can adapt for your own use – both in terms of the number of sections and the colours. When using the spinners in the classroom, ensure students give the paperclips a good flick, so that they spin rapidly before stopping in a place that could not be pre-determined – you can see how they are intended to be used in Chapter 2.

We have also put some worksheets on the website which you can adapt for your own use, although it is worth considering how much support in the form of printed tables and frequency diagrams your class needs. Generally, less-able classes need more support, so that they do not lose the point of the activity by having to focus on the mechanics of recording and organising data, while more-able classes should be challenged to think about how best to record and organise their results in order to use them in analysis.

Each chapter starts with a two-page guide to planning for the classroom, containing a summary of the activity or scenario, a lesson outline and a suggested plan for a week (three lessons) at each level. This is followed by the detail of the activities, including suggestions on how to do the experiments, collect data and record the results ready for discussion and analysis. Analysis is then covered at each level, including extensions of the activity or analysis for students who need additional challenge without necessarily needing to move on to a more advanced level. Finally, where appropriate we have made suggestions for relating the activity to life outside the classroom.

The first two activities *Fair game?* and *Which team will win?*, can be used at all three levels discussed in Part 1: introductory, intermediate and higher. *The dog ate my homework!* covers intermediate and higher levels, while *Choosing representatives* is aimed at higher level.

Because our resources and way of teaching may well be unfamiliar, here are some recommendations to ensure successful classroom delivery:

- DO include the experiment, unless your class has already done it recently, and still has the experimental results from that occasion.

- DON'T start by asking the class what they predict! Letting data speak for itself before expecting students to make predictions is crucial because they're unlikely to have any real sense of what to expect before they do the experiment. Predicting with no real idea of what to expect is just asking students to guess or, worse, try to work out what answer you want.

- DO spend time, when discussing the experimental data, clarifying the sample space (the list of possible outcomes). If there are two stages, with a choice between two alternatives at each stage, then there must be four outcomes; for instance, tossing two coins has two stages – first coin second coin, with a choice between Head and Tail at both stages – and so the sample space is HH, HT, TH and TT.

- DO use a frequency tree to represent data so that students see how the choices are structured. A narrative for each set of branches will help them to understand what the tree represents (see the introductory-level analysis of *Which team will win?* in Chapter 5 for an example).

- DON'T rush to formal analysis before the class has considered in detail what the data is telling them. They should note any extreme results, and see how using the average results for the class 'smooths' out variability. They are likely to have an intuitive grasp that more data brings the result closer to what might be expected, and this can be developed by observing the difference in variability of individual results and the whole class's results.

- DON'T rush the intermediate stage of putting expected frequencies on a frequency tree. Ensuring that all students fully understand the difference between what actually happened in their experiment and what we expect to happen is vitally important.

- DON'T rush moving from expected frequencies to probability trees. The process of calculating expected frequencies provides the basis for multiplying along the branches of a probability tree.

Fair game?

4.1 Planning for the classroom

Summary

Suitable for introductory, intermediate and higher levels.

In later chapters we use spinners as tools for mathematical modelling, but to start with we explore simple games based on the spinners. In each game, students' focus should be on the question 'Do you think this is a fair game?' meaning: does each player have an equal chance of winning? If yes, how do we know? If not, why not?

Lesson outline

Before the lesson, each group (pairs or threes as suggested) will need:

* spinners and worksheets as required.

5 mins at the most	Brief explanation and demonstration of how to play the game(s) – keep this as short as possible, with no questions at this stage about expectations.
10 mins	Students collect and record data.
10 mins	Students tally data and represent it on a frequency tree.
While students are completing tally and frequency tree	Teacher collates data from all groups and averages it, displaying the averages (1dp) on a results frequency tree on the board.
10 mins	Class discussion, leading up to 'Is this a fair game? Why (not)?'
As required	Analysis of the game at the appropriate level and comparison with other games if relevant.

Suggested plan for a week

Introductory level: students start with a single spinner, and then combine two spinners.

Intermediate level: students work with two sets of two spinners, and then a set of three spinners.

Higher level: students consider combined events empirically and theoretically, then generalise.

	Introductory level	Intermediate level	Higher level
Lesson 1	Game 1	Game 3	Games 3 and 4
Lesson 2	Game 2	Game 4	Games 3 and 5
Lesson 3	Game 3	Game 5	General result for Games 3 and 4, or Games 3 and 5
Notes	Display class average results (1dp) for each game, comparing with individual groups' results.	Compare class average results (1dp) with expected frequencies.	Compare empirical results, expected frequencies, and probabilities; use probability trees to investigate general results.
Extension	Compare expected frequency tree with averaged experimental results.	Spin just one of the unequal green/yellow spinners twice, and compare with Game 5.	Generalise across games.
Real-life context	There is potential to develop games like these for a STEM lesson looking at the mathematics behind genetic inheritance. In particular, Games 3 and 4 model Gregor Mendel's experiments between 1856 and 1863, in which he established the rules of heredity that lie behind the modern science of genetics. Visit www.nrich.maths.org/9270 for mathematical problems on genetic inheritance.		

4.2 The experiments

Game 1 (in pairs)

Figure 4.1 Blue/red spinners for Game 1

Each game consists of choosing one spinner from Figure 4.1, spinning it 12 times, and recording whether the outcome is red or blue. Player A wins if the outcome is red, Player B if it is blue. The game should then be repeated with the second spinner, and then with the third, giving three sets of 12 results.

> **Key questions**
>
> For each spinner, how many times did each player win?
> If you played for a very long time, what would you expect?
> Does it make a difference which spinner you are using? Why (not)?
> Are these fair spinners? Why (not)?

Game 2 (in pairs)

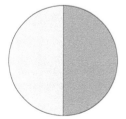 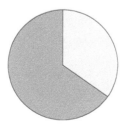 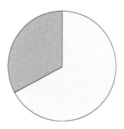

Figure 4.2 Yellow/green spinners for Game 2

Pairs start Game 2 with one of the spinners from Figure 4.2, spinning it 12 times, and recording whether the outcome is yellow or green. Player A wins if the outcome is yellow, Player B if it is green. Repeat with the second spinner, and then with the third.

> **Key questions**
>
> For each spinner, how many times did each player win?
> If you played for a very long time, what would you expect?
> Does it make a difference which spinner you are using? Why (not)?
> Are these fair spinners? Why (not)?

Game 3 (in pairs)

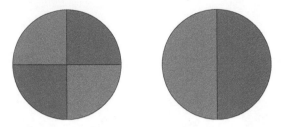

Figure 4.3 Blue/red spinners for Game 3

Pairs are given two blue/red spinners (Figure 4.3). They spin one and then the other, recording whether each is red or blue. If both are the same colour, Player A wins. If they are different, Player B wins. Groups should repeat this 24 times, unless they are going to compare with Game 5, in which case 36 trials will be needed (this is explained in the higher-level analysis later in the chapter).

> **Key questions**
>
> How many times did each player win?
> Were you surprised at your results? Why (not)?
> If you played for a very long time, what would you expect to happen?
> Is this a fair game? Why (not)?

Game 4 (in threes)

Figure 4.4 Blue/red spinners for Game 4

Groups are given three blue/red spinners (Figure 4.4). They spin each one in turn, recording the result of each spin. If all three are the same colour, Player A wins; if the result is two blue and one red, Player B wins; if the result is one blue and two red, Player C wins. Groups should repeat this 24 times.

> **Key questions**
>
> How many times did each player win?
> Were you surprised at your results? Why (not)?
> If you played for a very long time, what would you expect to happen?
> Is this a fair game? Why (not)?

Game 5 (in pairs)

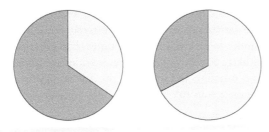

Figure 4.5 Green/yellow spinners for Game 5

Pairs are given two green/yellow spinners (Figure 4.5). They spin one and then the other, recording whether each is green or yellow. If both are the same colour, Player A wins. If they are different, Player B wins. Groups should repeat this 36 times.

> **Key questions**
>
> How many times did each player win?
> Were you surprised at your results? Why (not)?
> If you played for a very long time, what would you expect to happen?
> Is this a fair game? Why (not)?

4.3 Introductory-level analysis

The three spinners in Game 1 are all B : R = 1 : 1. While we might expect that blue and red will each occur half the time, individual results will, of course, vary from that. Calculating the class average provides a way of smoothing the data and removing the extremes, and thus giving a better indication of what we would expect to happen over a large number of games. It is a good idea to show the average to one decimal place, partly to avoid rounding errors and partly to reinforce that this is not raw data.

> **Key questions**
>
> Questions for class discussion could include:
>   Did anyone get exactly half red and half blue?
>   Why weren't all the results exactly 50–50?
>   Who has the most extreme results?
>   What was the longest run of red (blue)?
>   If a spinner gives considerably more red than blue (or vice versa), does this mean it is unfair? (No, it is not unfair, since red and blue are equally likely to occur on any spin, but we cannot predict whether any particular spin will be red or blue, and long runs of one colour will occur from time to time.)

In Game 2, one spinner is G : Y = 1 : 1, another is G : Y = 2 : 1, while the third is G : Y = 1 : 2. As with the blue/red spinners, some pairs may get results close to what might be expected, while others may show considerable variation from this. It is perfectly possible for the G : Y = 2 : 1 spinner to give yellow more times than green, even though green is twice as likely as yellow. This game provides an opportunity to talk about random variation and may well provide an opportunity to focus on the extreme results that may occur as a consequence.

For both Game 1 and Game 2, the experimental results can be displayed on simple frequency trees (Figure 4.6).

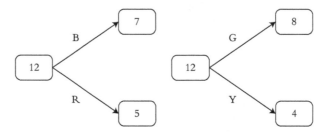

Figure 4.6 Exemplar experimental results frequency trees for Games 1 and 2

These simple frequency trees introduce a way to record data, and even at this early stage in students' experience of probability, they can discuss whether results like these are what we would expect, given the colour distribution on the spinner. The averaged class data can usefully be displayed next to an expected frequency tree for comparison (Figure 4.7).

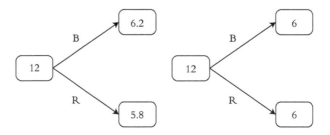

Figure 4.7 Exemplar averaged data and expected frequency trees for Game 1

Game 3 combines two of the blue/red spinners. For many students, this is the point at which probability becomes problematic. One cause of problems at this stage is seeing the possible outcomes as two blue, two red, and one of each. We avoid that by not discussing what might happen or what might be predicted before we have collected data and displayed it on a tree. The outcomes are then provided by the tree structure. Another reason is that students may continue to see two single spinners, rather than a pair. In an observed lesson, when the teacher was discussing whether combining two spinners would give a fair game, the students thought not, and when prompted said it was because the spinners individually were unfair – they had insufficient experience to be aware that a game combining two spinners will have different properties from a game with either spinner on its own. (The two spinners were those used in Game 5, so they were individually unfair and also unfair when combined. However, our expectation as teachers was that the students would think the two combined spinners 'compensated' for each other, making the game fair overall.)

In Game 3, the possible outcomes are BB, BR, RB and RR. Player A wins if both spinners are the same colour (BB or RR), and Player B wins if the spinners are different (BR or RB). Depending on time and how well the students are coping, it could work to show how the expected frequency tree for the two spinners is developed by adding the second spinner's tree at the end of each branch of the first tree, and adjusting the frequencies in the second tree branches accordingly.

4.4 Intermediate-level analysis

We use the tree structure for experimental results because familiarity with it will help students when they need to calculate what they expect to happen, given the spinners they are using (Figure 4.8).

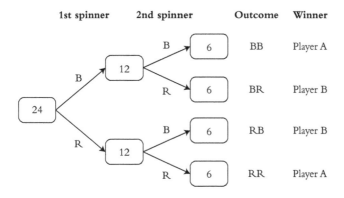

Figure 4.8 Expected frequency tree for Game 3

From the expected frequency tree, we can see that Game 3 is a fair game, because both players have 12 chances out of 24 to win.

When we come to Game 4, however, where we add in the third spinner, this changes what we expect to happen, and although each spinner individually has a 50–50 chance of giving blue or red, the combination of the three spinners no longer gives an even split. As we can see from Figure 4.9, there are now 8 outcomes – BBB, BBR, BRB, RBB, BRR, RBR, RRB, RRR – and three players, so an even split is impossible.

Player A wins if all three spinners show the same colour (outcomes BBB and RRR), so can expect to win on 6 occasions out of 24, or 25% of the time. Player B wins if the outcome is two blue and one red (BBR, BRB and RBB), so can expect to win on 9 occasions out of 24, or 37.5% of the time. Player C wins if the outcome is one blue and two red

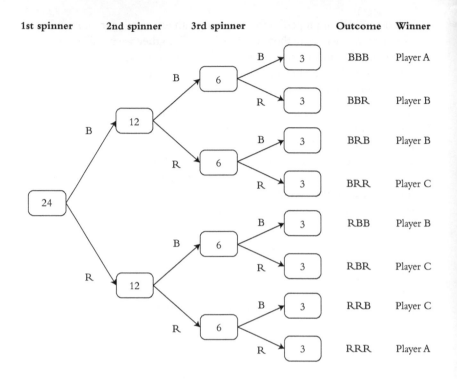

Figure 4.9 Expected frequency tree for Game 4

(BRR, RBR and RRB), so can also expect to win on 9 occasions out of 24, or 37.5% of the time. Game 4 is therefore not a fair game, as Player A has fewer chances to win than either Player B or Player C.

Combining the G : Y = 2 : 1 spinner with the G : Y = 1 : 2 spinner, intuition might suggest that they would balance each other, and that Game 5 would therefore be a fair game. The expected frequency tree (Figure 4.10) shows that this intuition would be incorrect. Player A has 16 chances out of 36 to win, compared to Player B's 20 chances out of 36.

The extension at this level, spinning the G : Y = 2 : 1 or the G : Y = 1 : 2 twice, is analysed at the end of this chapter

Figure 4.10 Expected frequency tree for Game 5

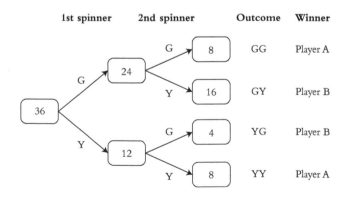

4.5 Higher-level analysis

Students should be encouraged to complete and discuss the experiments, and compare the averaged class results tree with the expected frequency tree, before moving on to probability trees, as comparing empirical and theoretical data helps students to avoid misconceptions based on faulty theorising. At this stage, they should consider why they have been instructed to do 24 trials for Games 3 and 4, but 36 when comparing Games 3 and 5 (they could use the 24 results for Game 3 in Lesson 1 and then just do another 12 trials).

Introducing the probability tree for the first time is an important step forward (see the discussion of this in the higher-level analysis of *Which team will win?* in Chapter 5). We emphasise that the numbers on the branches are probabilities not frequencies. Whereas with frequency trees we check that the sum of the frequencies at the ends of branches coming from a single node is equal to the frequency at the node, for probabilities we check that they sum to 1.

The probability tree for Game 3 (Figure 4.11) shows that each outcome is equally likely, with probability $\frac{1}{4}$, corresponding to the expected result of $\frac{6}{24}$ for each outcome. Players A and B are equally likely to win, since the probability of either winning is $\frac{1}{4} + \frac{1}{4} = \frac{1}{2}$.

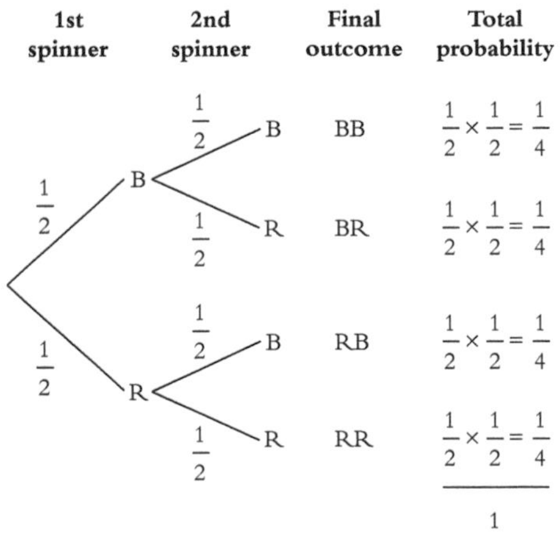

1st spinner	2nd spinner	Final outcome	Total probability
	$\frac{1}{2}$ B	BB	$\frac{1}{2} \times \frac{1}{2} = \frac{1}{4}$
$\frac{1}{2}$ B	$\frac{1}{2}$ R	BR	$\frac{1}{2} \times \frac{1}{2} = \frac{1}{4}$
$\frac{1}{2}$ R	$\frac{1}{2}$ B	RB	$\frac{1}{2} \times \frac{1}{2} = \frac{1}{4}$
	$\frac{1}{2}$ R	RR	$\frac{1}{2} \times \frac{1}{2} = \frac{1}{4}$
			1

Figure 4.11 Probability tree for Game 3

As we can see from Figure 4.12, the probability of each equally likely outcome for Game 4 is $\frac{1}{8}$. When Games 3 and 4 are compared, choosing 24 trials makes sense because we need the number of trials to be divisible by 4 for Game 3, and by 8 for Game 8. (Clearly 16 trials would also work mathematically, but doing more trials is always to be preferred if practical.)

Player A wins if the outcome is BBB or RRR, so the probability is $\frac{1}{4}$. Player 2 wins if the outcome is BBR, BRB or RBB, so the probability is $\frac{3}{8}$. Player C wins if the outcome is BRR, RBR or RRB, so the probability is also $\frac{3}{8}$. We see that Player A has only two chances in eight

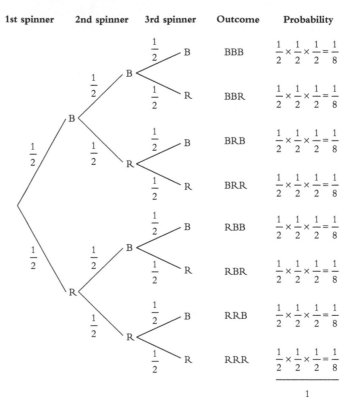

Figure 4.12 Probability tree for Game 4

to win, compared to the three chances in eight of Players B and C, confirming that this is not a fair game.

For Game 5 (Figure 4.13), the outcomes are not equally likely: $P(GG) = \frac{2}{9}$, $P(GY) = \frac{4}{9}$, $P(YG) = \frac{1}{9}$ and $P(YY) = \frac{2}{9}$. To compare Games 3 and 5 we therefore need a number of trials which is divisible by 4 for Game 3, and by 9 for Game 5 – hence 36.

Player A wins if the outcome is GG or YY, so the probability is $\frac{4}{9}$. Player B wins if the outcome is GY or YG, so the probability is $\frac{5}{9}$.

Figure 4.13 Probability tree for Game 5

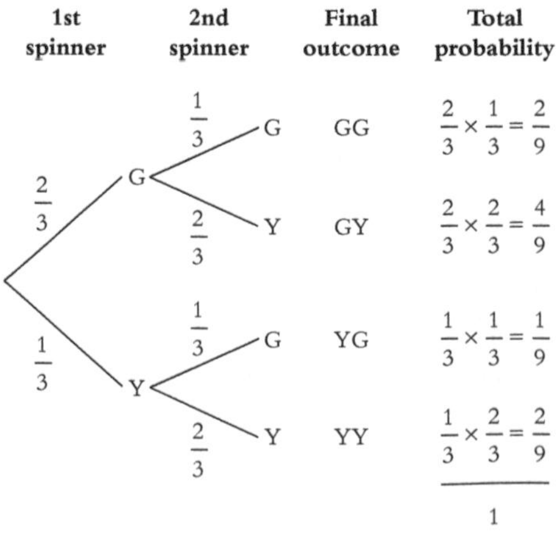

1st spinner	2nd spinner	Final outcome	Total probability
$\frac{2}{3}$ G	$\frac{1}{3}$ G	GG	$\frac{2}{3} \times \frac{1}{3} = \frac{2}{9}$
	$\frac{2}{3}$ Y	GY	$\frac{2}{3} \times \frac{2}{3} = \frac{4}{9}$
$\frac{1}{3}$ Y	$\frac{1}{3}$ G	YG	$\frac{1}{3} \times \frac{1}{3} = \frac{1}{9}$
	$\frac{2}{3}$ Y	YY	$\frac{1}{3} \times \frac{2}{3} = \frac{2}{9}$

1

Extending the analysis of Game 5

Although any of the games could be extended to general cases, we focus on Game 5. Instead of pairing $G : Y = 2 : 1$ with $Y : G = 1 : 2$, what happens if we pair either of these with itself? (This is the extension for the intermediate level.)

Figure 4.14 shows the tree for the $G : Y = 2 : 1$ spinner; the corresponding tree for the $G : Y = 1 : 2$ spinner is similar with the labels reversed. If Player A wins when the outcome is two colours the same (GG or YY), their probability of winning is $\frac{5}{9}$, compared to a probability of $\frac{4}{9}$ of Player B (GY or YG) winning.

One way to provide a challenge for students is to consider what happens if we generalise for $G : Y = p : 1 - p$ spun twice, as a single case. Instead of thinking about the spinners in terms of ratios of integers ($G : Y = 1 : 2$ for instance) we are now thinking of them in terms of ratios of probabilities, which allows us to generalise the analysis beyond what is possible with a physical spinner (see Figure 4.15).

The probability of the event 'matched colours' is $p^2 + (1 - p)^2$ and the probability of the event 'unmatched colours' is $2p(1 - p)$.

$p^2 + (1 - p)^2$ is a quadratic function which models Game 5 for values of p between 0 and 1. Graphing it will show that it has a minimum at $\left(\frac{1}{2}, \frac{1}{2}\right)$, so the probability of Player A winning (who wins when the spins match) is greater than $\frac{1}{2}$ for all other values of p. Clearly, if $p = 0$ or 1, then the probability of Player A winning must be 1, and the probability of Player B is 0. Games like this can only be fair, with both players having an equal chance of winning, when $p = \frac{1}{2}$ (corresponding to our blue/red spinners).

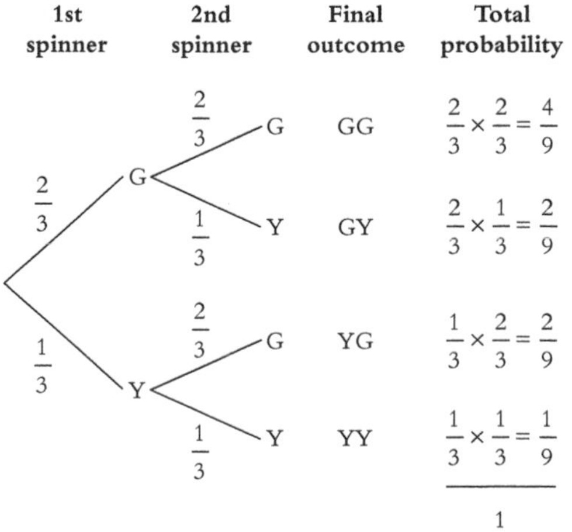

Figure 4.14 Probability tree for the $G : Y = 2 : 1$ spinner spun twice

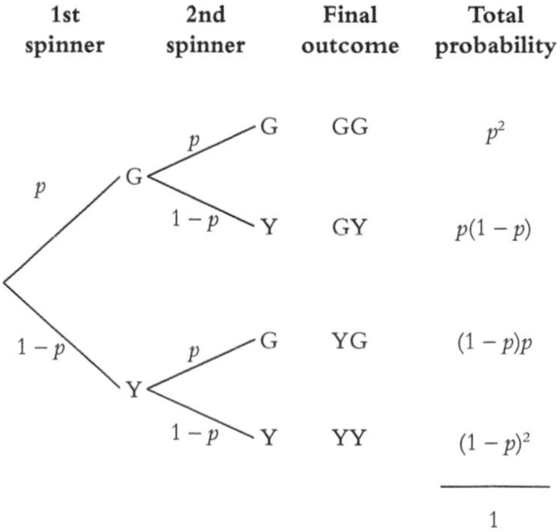

Figure 4.15 Probability tree for the general $G : Y = p : (1 - p)$ case

Which team will win?

5.1 Planning for the classroom

Summary

Suitable for introductory, intermediate and higher levels.

Every weekend, Team Yeti and Team Baboon play 2-goal football – they continue playing until two goals have been scored. Sometimes games take a long time, sometimes they are over in a flash, but either way they finish when two goals have been scored.

Team Baboon now have a new player! This gives them a huge advantage over their rivals. But how does their improved goal rate translate into games won? Who will win the season?

Lesson outline

Before the lesson, each group (threes suggested) will need:

- printed spinners
- worksheets as required.

5 mins at the most	Brief explanation and demonstration of how to play the game – keep this as short as possible, with no questions at this stage about expectations.
10 mins	Students collect and record data.
10 mins	Students tally data and represent it on a frequency tree.
While students are completing tally and frequency tree	Teacher collates data from all groups and averages it, displaying the averages (1dp) on a results frequency tree on the board. (Mini-whiteboards – if necessary made from plastic wallets with card inside them, which can be written on with a thick felt-tip pen – may make this easier.)
10 mins	Class discussion leading up to 'What are the chances of each team scoring a goal? What are their chances of winning the match?'
As required	Analysis of the game at the appropriate level.

Suggested plan for a week

All levels should play the game, collecting data for two 16-match seasons. This could be done at the introductory level, then revisited at intermediate or higher level for further analysis.

	Introductory level	Intermediate level	Higher level
Lesson 1	Collect and analyse data for 16 matches (Season 1) using Spinner 1.		
Lesson 2	Collect and analyse data for 16 matches (Season 2) using Spinner 1 and Spinner 2.		
Lesson 3	Compare results for the two seasons and complete analysis.		
Analysis	Display class average results (1dp) for each season, comparing with individual groups' results.	Compare class average results (1dp) for each season with expected frequencies.	Compare empirical results, expected frequencies, and probabilities; use probability trees to investigate general results.
Extension	Compare expected frequency tree with averaged experimental results.	Display experimental data and expected results on 2-way tables and/or Venn diagrams.	Generalise across spinners, perhaps extending to different numbers of goals.
Real-life context	Football matches are decided by many factors, but chance also plays an important part. Visit www.plus.maths.org/content/teacher-package-mathematics-sport#prediction for a range of articles on this topic.		

5.2 The experiments

Season 1 (in threes)

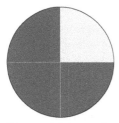

Figure 5.1 Spinner 1 for 2-goal football

Figure 5.1 shows the spinner we will use for Season 1. Demonstrate a few games with the whole class. Spin a paperclip around the centre of the chosen spinner. If it stops across a blue section, that represents a goal for the Baboons; if it stops across a yellow section, that represents a goal for the Yetis. Spin again for the second goal. Note who scored each goal, and what the result of the match was. Make sure that BY = 1–1 draw with the Baboons scoring first is distinguished from YB = 1–1 draw with the Yetis scoring first. Arguably they are different psychologically, and mathematically we need to differentiate them.

As soon as the class is happy with this, get groups to start collecting their data – groups of three work well for this activity. The outcome of each spin should be recorded, and then the result of each match. Some classes will be able to decide for themselves how to record their data, but others will need the support of a printed worksheet with a table for the experimental data and a tally table.

Resist the urge to ask the class what they expect, or what the chances of scoring or winning are, before they do the experiment! Allow the data to speak for itself first, and avoid putting students into a position where they are guessing or trying to read your mind. If any groups come up with interesting comments or questions during the experiment, for instance 'It's not fair! The Baboons have a lot more chances to score than the Yetis!' or 'The Baboons have 3 chances to score, the Yetis only have 1', simply record them on the board for discussion by the whole class.

It is a good idea for groups to have questions to answer as they finish, including one or two which will require discussion, so that no one is waiting idle while others catch up. Answering these questions will help students when they come to complete the frequency tree for their results.

> **Discussion questions**
>
> How many times did the Baboons win?
> How many times did the Yetis win?
> How many times did the Baboons score first, but the Yetis then equalised?
> How many times did the Yetis score first, but the Baboons then equalised?
> Do any of your results surprise you at all? Why (not)?
> The Baboons have more chances to score than the Yetis. What does this tell you about the teams?

Discuss the last two questions with the whole class once everyone has completed the experiment. Then collate the results from all the groups and display them on the board. Talk through the results: are there any particularly extreme sets of results? Then find the mean for each outcome to the nearest 1 decimal place. This will 'smooth' out the results, showing the underlying pattern.

> **Key questions**
>
> Other questions for whole-class discussion should include:
>   Why is this not a fair game?
>   How many chances do the Baboons have to score? How about the Yetis?
>   So what does it model? (Football games aren't fair when it comes to scoring and often one team is better than the other.)
>   Does the average number of wins for each team make sense to you, given the chances the Baboons have to score compared to the Yetis?
>   How does the frequency of draws compare with the frequency of wins?

The Baboons are three times more likely to score than the Yetis, but it is likely that in the averaged results the Baboons will have won around nine times more than the Yetis. One class of 11-year-olds talked about teams 'wasting some of their goals on draws', which stimulated a good discussion.

The number of YB draws should be roughly the same as the number of BY draws, which will surprise many students: it is not immediately obvious that 3 chances in 4 of scoring followed by 1 chance in 4 is the same as 1 chance in 4 followed by 3 chances in 4.

Season 2 (in threes)

Start the lesson in the same way by demonstrating the game for the students. The first goal will again be determined by Spinner 1 (Figure 5.1). If the Baboons score first, use Spinner 1 for the second goal also. However, if the Yetis score the first goal, use Spinner 2 (Figure 5.2) for the second goal. This changes the underlying model: if the Yetis score the first goal, then that improves their chances of scoring the second goal.

Students should collect data for a 16-match season again, recording their data in the same way as in the first experiment. The rest of the lesson then follows as before.

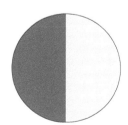

Figure 5.2 Spinner 2 for 2-goal football

5.3 Introductory-level analysis

Frequency trees make it much easier to understand the data. Before completing them, it will be helpful to answer the following questions.

> **Key questions**
>
> How many times did Y score the first goal?
> How many times did B score the first goal?
> How many times did Y score both goals?
> How many times did B score both goals?
> How many times did Y score first, then B equalised?
> How many times did B score first, then Y equalised?

The results for each season can then be recorded on frequency trees. A set of exemplar results for Season 1 is shown in Figure 5.3.

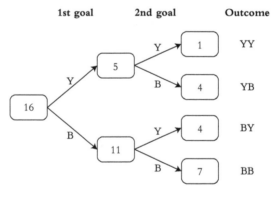

Figure 5.3 Exemplar results displayed on a frequency tree for Season 1

Students should be prompted to notice that the frequency preceding a pair of split branches is equal to the sum of the frequencies following the split: $16 = 5 + 11, 5 = 1 + 4, 11 = 4 + 7$, and then we also have $16 = 1 + 4 + 4 + 7$.

A frequency tree of the class results, averaged across the number of groups, should be displayed on the board. Spinner 1 gives B : Y = 3 : 1, while Spinner 2 gives B : Y = 1 : 1, but these ratios simply tell us each team's chance of scoring a goal. Students should observe that for Season 1 (using just Spinner 1), the Baboons win something like 9 times as often as the Yetis, while for Season 2, that should have dropped to nearer 4 or 5 times. It is really important that the teacher is very careful not to confuse the chance of scoring a goal with the chance of winning the match.

Each set of branches corresponds to a story or narrative for a match:

Top set of branches (YY): *The Yetis start today as the underdogs, but they're off to a great start – oh, yes, what a goal! Can they keep it up? Yes, they're on fire today – what a shot! Yes, it's a goal! A well-deserved win for the Yetis.*

Second set of branches (YB): *The Yetis start today as the underdogs, but they're off to a great start — oh, yes, what a goal! Can they keep it up? No, it's not looking so good for them … Oh, look at the Baboons' new player, what a shot! Yes, it's a goal, they've equalised. It's a draw.*

Third set of branches (BY): *And the Baboons are off to a great start — yes, they've scored! I see the Yeti's manager giving them a pep talk — let's see if it makes a difference. Oh, yes, look at them go now! Yes, they've scored the equaliser. It's a draw!*

Bottom set of branches (BB): *And the Baboons are off to a great start — yes, they've scored! Oh, and again! Wow, what a match! What a difference that new player is making!*

In any one match, only one of these stories will happen. Over a season, they will probably all happen, but of course they are not equally likely.

5.4 Intermediate-level analysis

To understand the results better, we need to look at the expected results. What should we expect to happen over a 16-match season?

For each goal, we need to know the chance that a team scores. It makes a difference how this is phrased. A student who talked of the Yetis scoring 25% of the goals, and the Baboons scoring 75%, found it much easier to work out what should go in the boxes on the expected frequency tree than others who focused on 1 chance in 4 or 3 chances in 4. In 16 matches, there are 16 first goals, so in Season 1 we expect the Yetis to score 4 of them, and the Baboons 12. For the 4 matches where we expect the Yetis to score the first goal, we expect them to go on to score the second goal 25% of the time, so once in those 4 matches, and in the other 3 matches we expect the Baboons to equalise. For the 12 matches where we expect the Baboons to score first, we expect the Yetis to equalise in 25% of them, so 3 times, and the Baboons to go on to win the other 9 matches (see Figure 5.4).

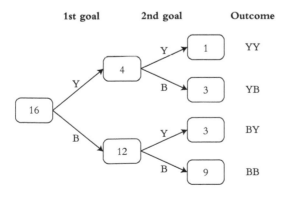

Figure 5.4 Expected frequency tree for Season 1

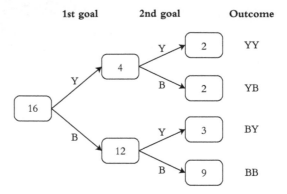

1st goal	2nd goal	Outcome

Figure 5.5 Expected frequency tree for Season 2

The same kind of reasoning also applies to Season 2, but we have to take into account the change to Spinner 2 if the Yetis score the first goal. This is shown in Figure 5.5.

Individual groups' results for Season 2 may not look very different from those for Season 1, and indeed some groups may find that the Yetis won more games in Season 1 than in Season 2. However, the averaged class results should show the Yetis doing better in Season 2, as would be expected with an improved goal rate in some matches at least.

> **Key observations**
>
> The averaged experimental results should be very close to the expected results, especially if the average has been rounded to the nearest 1 dp.
> For each pair of branches, the total is equal to the value at the branch point – so for Season 1 (Figure 5.4) 16 = 4 + 12, 4 = 1 + 3, 12 = 3 + 9, and also 16 = 1 + 3 + 3 + 9.
> The expected total number of draws in Season 1 is 3 + 3 = 6 and in Season 2 is 2 + 3 = 5. Adding the results for the sets of branches which lead to YB or BY seems trivial when we are dealing with whole numbers, but will not be so obvious when working with probability trees.
> In Season 1 the number of expected YB draws is equal to the number of expected BY draws. Do the students feel this is what they would expect? Can they explain it? What has changed in Season 2?

From numbers to proportions

Take the students through a series of questions. The answers should be taken directly from the expected frequency trees, and are designed to help students understand more fully what the tree is telling them.

Key questions

What proportion of the matches do you expect the Yetis/Baboons to win?

What proportion of the matches do you expect the Yetis to score first, then the Baboons to equalise?

What proportion of the matches do you expect the Baboons to score first, then the Yetis to equalise?

How do these proportions compare with those for the experimental results?

In Image 5.1, you can see how one teacher helped students to see that although they knew the Baboons were three times as likely to score as the Yetis, in their averaged experimental results (shown here to the nearest whole number) the ratio of wins was 8 : 1, while for the expected results the ratio is 9 : 1.

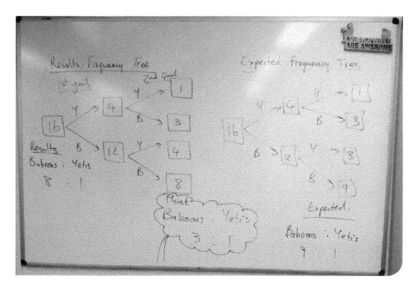

Image 5.1 What do you expect to happen?

Extending the analysis

Once students are comfortable with representing their data on frequency trees and using the trees to work out what the expected results would be, they can be encouraged to represent their data and/or the expected results in 2-way tables (Table 5.1) or Venn diagrams (Figure 5.6).

Labelling the circles in a Venn diagram is not trivial. The easiest way to do it is to write down the sample space (the list of all possible outcomes, not just those which actually happened in an experiment), then to assign one to each region, ensuring that the overlaps make sense. For the football game, the outcomes are YY, YB, BY and BB. We could label the

		2nd goal		Total
		B	Y	
1st goal	B	9	3	12
	Y	3	1	4
Total		12	4	16

Table 5.1 2-way table for Season 1

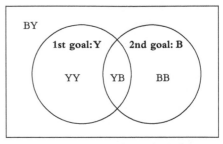

Figure 5.6 Venn diagram for the football game

intersection of the two circles YB, say, then put YY on the left of that – so the left-hand circle contains the number of matches in which the Yetis score first – and BB on the right – so that the right-hand circle contains the number of matches in which the Baboons score second. That leaves BY for the region outside the circles.

Finally we can insert the frequencies for each outcome as shown in Figure 5.7.

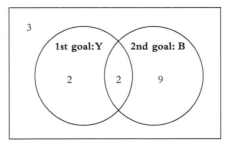

Figure 5.7 Venn diagram showing the expected results for Season 2

This is not simply an exercise in representing data: it is an opportunity to investigate what information each representation shows, what is implicit, and what is emphasised.

> **Key observations**
>
> All three ways of representing the experimental data actually contain the same information, but because it is displayed differently, attention is drawn to different features.
> The frequency tree emphasises the dynamic and narrative aspect – first one goal, then the second – with each set of branches providing a unique story (outcome).
> The 2-way table emphasises the totals, and the symmetry is perhaps more obvious.
> The Venn diagram emphasises the number of times each outcome occurs, with the other information implicit.

5.5 Higher-level analysis

This is a good activity for motivating the transition from frequency trees to probability trees, and also for discussing the difference between independent and dependent probabilities.

Start by working through the calculations required to complete the expected frequency tree, at each stage emphasising the process. An idealised classroom dialogue for Season 1 might go something like this:

Teacher: What is the chance that the Yetis score a goal?

Students: 1 out of 4, 25%.

T: Right, so in a season of 16 matches, how many times do we expect the Yetis to score the first goal?

S: 4 out of 16.

T: Good. So because just one quarter of the spinner is yellow, meaning the Yetis score, we expect them to score the first goal a quarter of the time, so that's four times in 16 matches.

How about the Baboons? In what proportion, what fraction, of the matches do we expect them to score the first goal?

S: 75%, three quarters.

T: And what's three quarters of 16?

S: 12, 3 × 4.

T: Yes, very good. We expect the Yetis to score the first goal in a quarter of the matches, and the Baboons to score first three quarters of the time. So we can say that the *probability* that the Yetis score the first goal is $\frac{1}{4}$ and the *probability* that the Baboons score the first goal is $\frac{3}{4}$.

When we move on to the second goal, are the probabilities the same? How do we know?

S: Yes, they are, because we're still using the same spinner, so it's still 1 chance out of 4 for the Yetis and 3 chances out of 4 for the Baboons.

T: Ok, good. What do the top branches on the expected frequency tree tell us about?

S: The number of times the Yetis score the first goal, and then the second, so they win.

T: Ok, and we found a quarter of 16, which is 4, then a quarter of that, which is 1. Changing that into probabilities, we want a quarter of a quarter. How do we calculate a quarter of a quarter?

S: Multiply.

And so on. This is an idealised dialogue, but it provides a rationale for why we multiply along the branches of a probability tree, based on finding a proportion of a proportion on the expected frequency tree. It is important to emphasise that working with the probabilities enables us to consider any number of matches. In general, we expect the Yetis to score a quarter of the first goals and a quarter of the second goals, so we expect them to win a quarter of a quarter of the time. Mathematically, that's $\frac{1}{4}$ of $\frac{1}{4}$ or $\frac{1}{4} \times \frac{1}{4} = \frac{1}{16}$.

A dialogue like this needs to be supported with frequent reference to both the expected frequency trees (Figures 5.4 and 5.5) and the probability trees (Figures 5.8 and 5.9).

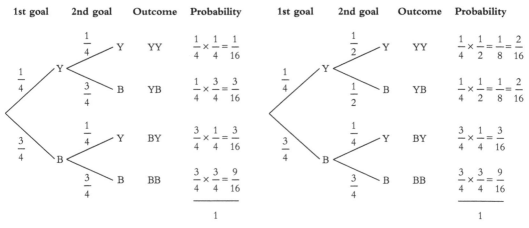

Figure 5.8 Probability tree for Season 1   Figure 5.9 Probability tree for Season 2

We have now reached the point where conventionally probability teaching starts – drawing a probability tree, identifying the outcomes for each set of branches, and then calculating the probabilities. The difference is that we have provided a scaffold to help students reach this point through analysing data to see what actually happens, then thinking about what we would expect to happen, and finally generalising to probabilities.

> **Key questions**
>
> In Season 1, does the second goal depend on the first in any way? How about in Season 2?

The difference between the two seasons, whether the same spinner is used for both goals or not, should prompt a discussion about independence and dependence (see Section 2.6 *The theory of probability* in Chapter 2).

Extending the analysis

Students should be reminded that this is a mathematical model of a football game. How does it compare with reality? How could we improve it, to make it a bit more realistic? Suggestions might include:

- Using different spinners to change the chance of a team scoring
- Changing the number of goals. It is perfectly possible to model different numbers of goals using tree diagrams, but in practice few will want to go beyond two or three goals! However, analysing the model for 0, 1, 2 and 3 goals using tree diagrams should give a good group enough data for them to generalise for n goals.

A further challenge would then be to generalise for n goals and the probability, p, that one team scores.

The dog ate my homework!

6.1 Planning for the classroom

Summary

Suitable for intermediate and higher levels.

A certain teacher, Mr L I Detector, claims that he can tell when students are lying when they make an excuse about their missing homework. This claim is true. Unfortunately, he also accuses some students who are telling the truth.

What are the chances that a truthful student is accused?

What are the chances that a student who is accused is actually telling the truth?

Lesson outline

Before the lesson, each group (threes or fours suggested) will need:

*   multi-link cubes in four different colours – provide at least 20 black, 15 green, 9 white and 4 orange (or equivalent)
*   spinners and worksheets as required.

5 mins at the most	Brief explanation and demonstration of data collection – keep this as short as possible, with no questions at this stage about expectations.
10 mins	Students collect physical data.
10 mins	Students tally data and represent it on a frequency tree.
While students are completing tally and frequency tree	Teacher collects (physical) data from all groups, displaying it on a large 2-way table and/or Venn diagram and/or frequency tree.
10 mins	Class discussion leading up to 'What are the chances that Mr D gets it wrong?'
As required	Analysis of the game at the appropriate level.

Suggested plan for a week

All groups will need to collect data for a class of 24 students. This could be done at the intermediate level, then revisited at higher level for further analysis.

	Intermediate level	Higher level
Lesson 1	Collect and analyse data for 24 students in Class X.	
Lesson 2	Collect and analyse data for 24 students in Class Y.	
Lesson 3	Compare results for the two classes and complete analysis.	
Analysis	Collect (physical) data from all groups for class display. Use expected frequency trees to analyse.	Collect (physical) data from all groups for class display. Use expected frequency trees and probability trees to analyse.
Extension	Display experimental data and expected results on 2-way tables and/or Venn diagrams; compare frequency trees.	Explore conditional probability in this experiment and more widely.
Real-life context	This scenario models many real-life contexts, such as testing people for illness or illegal drugs. Misunderstanding of conditional probability has led to some (in)famous miscarriages of justice, so ensuring students have some understanding of it is vital.	
	Visit www.motivate.maths.org/content/MathsHealth/testresults for video clips and classroom activities on this topic.	

6.2 The experiments

In the interests of clarity we have chosen specific colours for each event in the experiment, but of course they are quite arbitrary. The worksheets available on the website include a grey-scale version which can be adapted for other colours. If you do not have sufficient multi-link cubes in four colours, we have successfully used two different colours of bottle tops with two different colours of beans placed inside the bottle tops instead (see Image 6.4). The expected results for Class X require 20 black, 15 green, 9 white and 4 orange cubes per group of students, and for Class Y 20 black, 18 green, 6 white and 4 orange cubes are required.

The black/orange spinner in Figure 6.1 determines whether a student is telling the truth or not – black means they are truthful, orange means they are not. The green/white one determines whether, if they are telling the truth, Mr D accuses them – green means he believes them, white means he accuses them of lying.

Figure 6.1 Spinners for Class X and Class Y

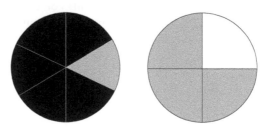

Class X (in threes or fours)

Start by demonstrating how to collect data. Data is collected physically as pairs of cubes (Image 6.1), or beans placed in bottle tops (Image 6.4).

Image 6.1 Data collection for *The dog ate my homework!*

Spin the black/orange spinner:

- black means the student is telling the truth – take a black cube
- orange means the student is lying – take an orange cube
- if the student is lying, Mr D will accuse them – stick a white cube onto the orange one.

If the student was telling the truth, spin the green/white spinner:

- green means Mr D believes the student – stick a green cube onto the black one
- white means Mr D thinks the student is lying – stick a white cube onto the black one.

Key questions

What does orange and white mean?
How about black and green, or black and white?
Why is orange and green not possible?
What is the sample space for this experiment?

Class Y (in threes or fours)

Mr D realises that his judgements are not always correct! How does it change the outcomes for the students if he does not automatically accuse people he suspects of lying?

Use the black/orange spinner again to determine if students are lying, then use the green/white spinner for all students to determine if Mr D accuses them or not.

Key questions

What are the possible outcomes now? So what is the sample space for this experiment?
Are your results for Class Y different from those for Class X or not?

6.3 Intermediate-level analysis

Groups should tally their data, recording it on a frequency tree (see Figure 6.2 for the structure). Meanwhile, set up a large 2-way table somewhere the class can gather around it, and invite groups to place their physical data on the table once they have recorded it (see Image 6.2 and Image 6.3). While they wait for other groups to finish, they should discuss the questions below, which can then be opened up for whole-class discussion around the collated data display.

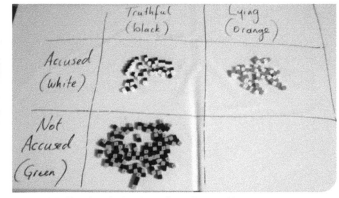

Image 6.2 Adding data to the large 2-way table

Image 6.3 Class data displayed on a large 2-way table

Using physical data on large displays is an alternative to averaging groups' data to see the effect of gathering more data. Students will probably be surprised to see that the number of students who are wrongly accused is similar to the number who are correctly accused when all groups' cubes are displayed together, even if that did not happen in individual groups' results.

The expected frequency tree (Figure 6.2) will help students to understand the data better. The black/orange spinner has 1 chance in 6 of giving orange, so of our 24 students we expect that 4 will be lying and 20 will be truthful. In Class X, all four of the lying students will then be accused by Mr D. To see what happens to the 20 truthful students, we use the green/white spinner. The chance of white (meaning accused) is 1 in 4 or 25%, so we expect that 5 of the 20 will be wrongly accused by Mr D, and the other 15 will not be accused. We expect that 9 students in total will be accused by Mr D, 4 correctly and 5 incorrectly.

Comparing what we expect for Class Y (Figure 6.3), we see that fewer students in total are likely to be accused, but that most of them are wrongly accused.

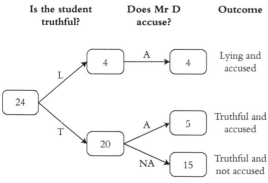

Figure 6.2 Expected frequency tree for Class X

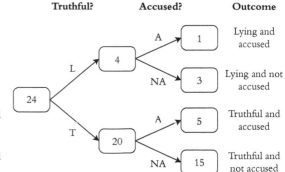

Figure 6.3 Expected frequency tree for Class Y

Extending the analysis

Large class displays of physical data can be used to compare data displayed on frequency trees, 2-way tables and Venn diagrams, as Image 6.4 illustrates.

So far, we have looked at the accused students in relation to the whole class of 24 students. We can also consider the unfortunate wrongly accused students in two further ways:

> **Key questions**
>
> What proportion of the truthful students are accused?
> What proportion of the accused students are truthful?

These two questions are not the same. In each case, we first need to identify the reference class. In the first question, we are considering the accused students who are a subset of those who are truthful; in the second question, we are considering the truthful students who are a subset of those who are accused. Both questions refer to the same group of students (who were represented by black and white cubes in the experiments), but they are now considered in relation to different subsets of the class of 24.

For Class X, we expect there to be 20 truthful students, of whom 5 ($\frac{5}{20} = 25\%$) are wrongly accused. We expect a total of 9 students to be accused, of whom 5 ($\frac{5}{9}$ is over 50%) are innocent. For Class Y, the analysis for the 20 truthful students is the same. However, we now expect only 6 students to be accused by Mr D, of whom 5 are actually innocent – so only $\frac{1}{6}$ or 17% of those accused are actually guilty!

Displaying the physical data (Image 6.4) at the ends of the branches of large frequency trees (Figure 6.4) will help students to understand the significance of the order in which the branches of a tree are presented, and to realise the importance of being very clear about the reference class when asking and answering probability questions.

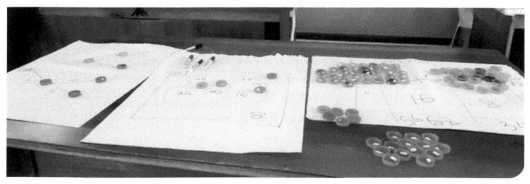

Image 6.4 Frequency tree, Venn diagram and 2-way table for *The dog ate my homework!* showing use of bottle tops and beans

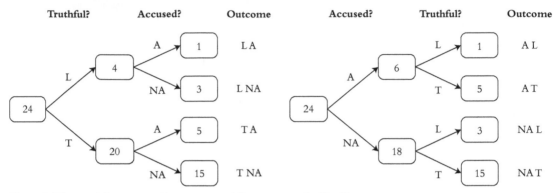

Figure 6.4 Expected frequency and inverse expected frequency trees for Class Y

We are laying the foundations for understanding conditional probability with these frequency trees. The analysis presented in this section highlights the importance of identifying the correct reference class, which depends on the precise wording of the question. We saw that the reference class for one question was the truthful students, while for the second it was the accused students. The answers to these two questions are therefore not the same, even though both concern the same group of five wrongly accused students. This is the difference between the chance that B occurs given that A has already occurred, and the chance that A occurs given that B has already occurred (this is further discussed in Chapter 22: *How should we change our beliefs?* and Chapter 24: *Misconceptions*).

6.4 Higher-level analysis

This activity can be used to help students with the transition from frequency trees to probability trees. In the higher-level analysis of *Which team will win?*, there is a description of how modelling a football game can be used to motivate the transition from frequency trees to probability trees, which could be adapted for use with this activity. This activity is also useful for prompting discussion about independent and dependent events, as for Class Y the accusations are entirely independent of guilt!

In Class X, five-sixths of the students tell the truth, but a quarter of those will be wrongly accused. This means taking $\frac{5}{6}$ of the 24, and then finding a quarter of the answer. Mathematically this is $\frac{5}{6} \times \frac{1}{4}$ (see Figure 6.5).

This can be compared with the probability tree for Class Y (Figure 6.6).

The key questions and discussion in the intermediate-level analysis can also be used at this level, but using probability trees in addition to expected frequency trees.

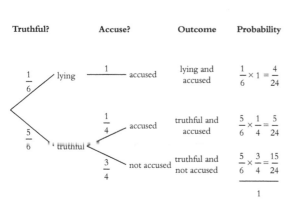

Figure 6.5 Probability tree for Class X

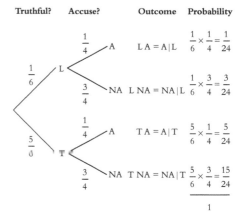

Figure 6.6 Probability tree for Class Y

Extending the analysis

We have seen what difference it makes if Mr D knows that he is not always correct when he accuses students of lying. Students could be challenged to develop the model further, by identifying and critiquing the assumptions upon which it rests. How could it be made more realistic? (One class observed that the chances of someone lying about their homework were much greater than 1 in 6 – 'My printer didn't work', 'I'll bring it tomorrow/after lunch', … – and that the chances of the teacher accepting what they said at face value were close to 1!)

This problem models, in a humorous way, the mathematics behind many tests, such as testing DNA found at a crime scene and testing for cancer, pregnancy or illegal drugs. Tests are not 100% perfect – there will be occasions when a positive result is incorrect, and also occasions when a negative result is incorrect. This is discussed further in Chapter 1, and also in Chapter 22: *How should we change our beliefs?*

This problem can also be used to discuss conditional probability in an informal way, as we suggest in the questions about the accused but innocent students. Given that a student is truthful, what is the probability that they will be wrongly accused? Given that a student is accused, what is the probability that they will in fact be innocent? Such questions are much easier to answer from the frequency tree than from the probability tree, which is one of the reasons why we advocate the use of frequency trees so strongly.

This may seem irrelevant for students who are not studying more advanced mathematics, but the failure to understand conditional probability lies at the root of several miscarriages of justice. The news that a test for HIV or cancer is positive is devastating – how much more so if it is later found to be incorrect? In our opinion, understanding conditional probability is vital for us all.

Choosing representatives

7.1 Planning for the classroom

Summary

Suitable for higher level.

A group of eight students, five girls and three boys, entered a business enterprise competition. They were thrilled when they won, and two of them were invited to represent the group at an award ceremony with other prize-winners and entrepreneurs. But how were they to choose the two?

Emily and Daniel claimed they did most of the work, so it should be them. The others disagreed, and came up with three ways to choose their representatives. Which method do you think is best? Which would work best for choosing representatives from a much larger number?

Lesson outline

Before the lesson, each group (twos or threes suggested) will need:

* counters or cubes – four in colour 1, two in colour 2, and one each in two other colours

* two opaque bags or boxes to draw counters or cubes from.

5 mins at the most	Brief explanation and demonstration of data collection – keep this as short as possible, with no questions at this stage about expectations.
10 mins	Students do the experiment.
10 mins	Students tally data and represent it on a frequency tree.
While students are completing tally and frequency tree	Teacher collates data from all groups, averages it (1dp) and displays it on a frequency tree for class discussion.
10 mins	Class discussion leading up to 'What are the pros and cons of this method of choosing representatives?'
As required	Analysis of the game at the appropriate level.

Suggested plan for a week

Three methods of choosing representatives are given below. Each could form the main activity for a lesson, along with the analysis of it, although in the third lesson time should also be allocated for comparison of the three methods.

	Higher-level
Lesson 1	Lucia's method – collect and record experimental data, analyse expected results.
Lesson 2	Maria's method – collect and record experimental data, analyse expected results.
Lesson 3	Jacob's method – collect and record experimental data, analyse expected results. Compare the three methods.
Analysis	Collated data averaged and displayed on a frequency tree for class discussion. Probability trees used to analyse expected results.
Extension	Analyse methods from the perspective of a specific girl (Emily) and boy (Daniel).
Real-life context	This scenario introduces students to the difficulties in devising a fair system for choosing eg. elected members of a parliament or assembly. Scale matters – the method they think works best for eight people may be different from the method they would prefer for a large number of people.
	Visit www.plus.maths.org/content/taxonomy/term/855 for a range of articles on this topic.

7.2 The experiments (in twos or threes)

Start by assigning identities to the counters. The four in colour 1 represent the girls apart from Emily; choose one of the single counters to represent her. The two in colour 2 represent the boys apart from Daniel; the other single counter represents him.

Each experiment consists of 16 trials, with the results of the first and second choices recorded for each trial – girl or boy – and whether Emily or Daniel are included or not. Students should then tally their results for the sample space GG, GB, BG and BB, so not distinguishing Emily and Daniel at this stage, and display their experimental data on a frequency tree, comparing their own results with the averaged results for the whole class. In Image 7.1, you can see two examples of students' work – by this stage, they should be capable of organising their own methods for recording data.

Image 7.1 Recording experimental results and frequency tree

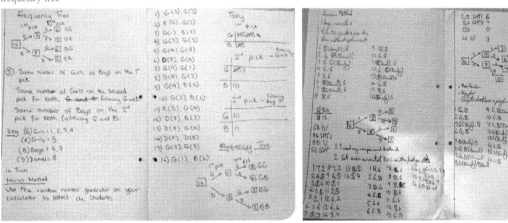

Lucia's method

Put everyone's name on pieces of paper, and fold them over so the names cannot be seen. Put the pieces of paper in a bag, mix them up, and pick out two.

Model this by putting all the counters into one opaque bag or box, then pick out two counters, one at a time, without replacing the first before you take out the second.

Maria's method

Give everyone a number from 1 to 8. Then use the random number generator on a calculator to choose two people.

Model this by numbering the group from 1 to 8 – for instance,
1 = Emily, 2 = Girl, and so on. Then use =RANDBETWEEN(1,8)
in Excel or the random integer function on a calculator to generate
random numbers between 1 and 8. Numbers should be recorded in pairs,
with repetitions not excluded – these should be recorded as valid results
for this method.

Jacob's method

Put the boys' names on blue cards, and the girls' names on red cards. Pick one card
at random from the girls' cards, and one from the boys' cards.

Model this by putting the girls' counters (including Emily's) in one opaque
bag or box, and the boys' counters (including Daniel's) in the second. Take
one counter out of each bag or box, and note down the results.

Image 7.2 shows the results of one class, recorded on frequency trees.
Students then discussed the results and the methods.

Image 7.2 Frequency
trees for the three
methods of choosing
representatives

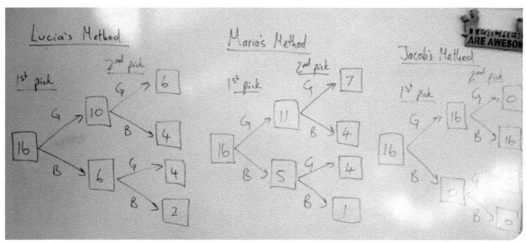

Key questions

What proportion of the 16 trials were for each outcome (GG, GB,
BG, BB)?
Compare your results to the class average results. Are you surprised
by any of them?
Which method do you think was fairest? Why?
Which method would you choose at this stage if you were Emily?
Why?
How about Daniel? Why?

7.3 Higher-level analysis

The first method (Lucia's) requires the group to choose two people from a group of eight, without replacing the first choice before making the second. The second method (Maria's) is very similar, except that this time the first choice is replaced after being chosen, so could be chosen twice, and while this clearly has implications for choosing two names from a group of eight, comparing these two experiments should help students to understand the difference between sampling with and without replacement. The third method (Jacob's) involves a form of stratification, with one representative chosen from the girls and one from the boys.

Once students have completed each experiment, and recorded and commented on the experimental results, they should complete a probability tree for each method (see Figure 7.1, Figure 7.2 and Figure 7.3).

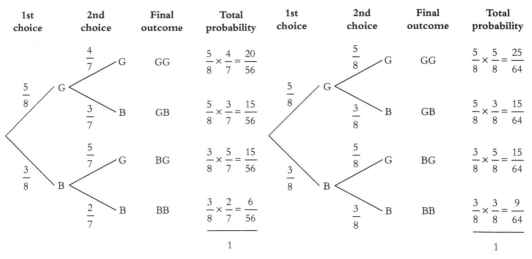

Figure 7.1 Lucia's method: sampling without replacement

Figure 7.2 Maria's method: sampling with replacement

For Jacob's method, we have chosen to pick the girl first – picking the boy first simply reverses the two stages. The probability of picking a girl from a sample of girls is of course 1, and the probability of picking a boy is 0. For the second choice, from a sample of boys, it is certain that a boy will be picked. Some students will see that there is no real need to draw a tree here, but others may not be clear about representing probabilities of 0 and 1 on trees, so this is an opportunity to discuss this.

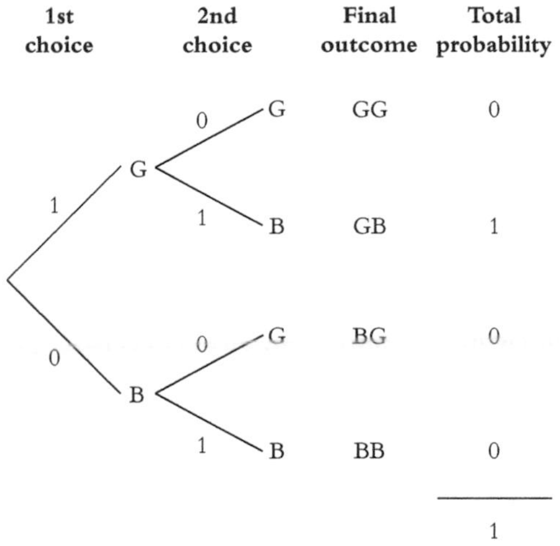

1st choice	2nd choice	Final outcome	Total probability

Figure 7.3 Jacob's method: stratified sampling

> **Key questions**
>
> Which do you think provides the best way for the group to choose their representatives? Why?
> Is this also the method you thought best from the experimental data? Why (not)?
> Can you think of an alternative method which might be better than these?

There is no one right answer to the question as to which method is best (unless you are Emily or Daniel!). Most of the students trialling these activities felt that Lucia's method (random sampling without replacement) was best for a small number. This view was typical: 'I think Lucia's is the most fair as there is no bias because it is completely random and you cannot be chosen twice.' Opinion was divided on Maria's method: 'I think Maria's method is the most fair because a machine is generating numbers randomly and everyone has the same chance to be picked twice in a row', wrote one student, while another thought it 'less reliable … as it is possible to get the same person twice.'

Students should also consider what difference scale makes. One student observed that: 'More girls were picked than boys in both 1st and 2nd [method].' Jacob's method (stratified sampling) was therefore considered better for a large number because it ensures subgroups are fairly represented. Another student felt that Maria's method would work well for a large group, because the chance of someone being chosen twice would be much smaller.

Extending the analysis

Students should construct the probability trees from Emily's and Daniel's perspective for each method. The first question to ask is whether the tree is the same for both of them, because if it is then only one needs to be constructed.

For Lucia's method, the tree will be the same for either, since Lucia's method does not distinguish between boys and girls.

As we can see in Figure 7.4, both Emily and Daniel have a 1 in 8 chance of being chosen the first time, and a 1 in 7 chance the second time. If they are chosen the first time, then we do not need to consider what might happen in the second choice. The probability of their being chosen either first time or second time is $\frac{1}{8} + \frac{1}{8} = \frac{1}{4}$.

Figure 7.5 shows part of an alternative analysis, submitted for homework by a Year 10 student.

Maria's method (see Figure 7.6) does not distinguish between girls and boys either – there is a 1 in 8 chance of being chosen the first time, and also a 1 in 8 chance of being chosen the second time.

From the tree we see that the probability of being chosen either time is $\frac{8}{64} + \frac{7}{64} = \frac{15}{64}$, which is less than the $\frac{16}{64} = \frac{1}{4}$ for sampling without replacement (Lucia's method). This is to be expected, since in Maria's method there are 8 names to choose from, rather than 7, for the second choice.

For Jacob's method (Figure 7.7), the analysis for Emily and Daniel is different, because it does differentiate between girls and boys. Emily has a 1 in 5 chance of being chosen as the girl; Daniel has a 1 in 3 chance of being chosen as the boy.

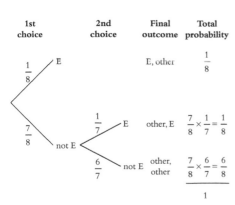

Figure 7.4 Lucia's method from Emily's (or Daniel's) perspective

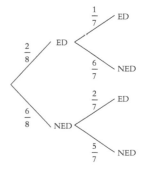

Key:

ED – Emily or Daniel

NED – Not Emily or Daniel

Outcome	Probability
ED ED	$\frac{1}{28}$
ED NED	$\frac{3}{14}$
NED ED	$\frac{3}{14}$
NED NED	$\frac{15}{28}$

Figure 7.5 Lucia's method from Emily's and Daniel's perspective

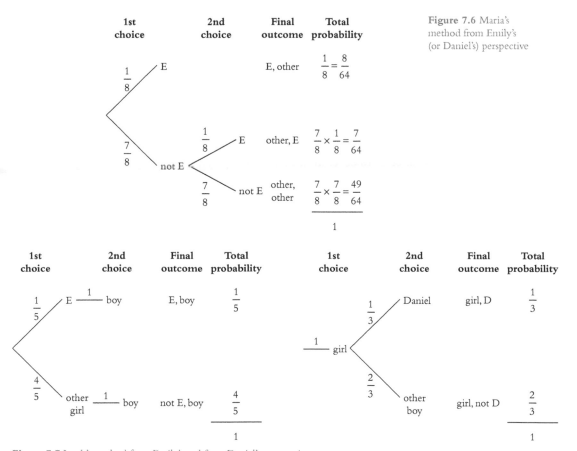

Figure 7.6 Maria's method from Emily's (or Daniel's) perspective

Figure 7.7 Jacob's method from Emily's and from Daniel's perspective

Clearly, Daniel would prefer Jacob's model, as this gives the best chance for him to be chosen $\left(\frac{1}{3}\right)$. Emily, on the other hand, would prefer Lucia's method, since the 1 in 4 chance of being chosen is her best bet.

7.4 Real-life context

Just as there is no one perfect way to choose representatives from a group of students, so there is no perfect way to choose the people who represent us in national or local government, which is why it is a contentious issue. First-past-the-post provides clear-cut winners, but often means that a so-called majority government is in fact a minority government. In 2015, the British people elected a majority Conservative government: the Conservatives won 330 seats out of 650, so were the largest party in the House of Commons. On the other hand, their share of the votes cast was 37%, with only 24% of eligible voters voting Conservative. What might the House of Commons have looked like had there been some form of proportional representation? We'll never know, but it certainly would have been different!

Introduction

Part 3 features a wide collection of sample assessment questions, based on similar questions set in past GCSE examinations and those proposed (in 2015) for the revised Mathematics GCSE at both Foundation and Higher tiers in England and Wales. For each question we have indicated both the tier and where we think questions sit in our curriculum, with its introductory, intermediate and higher levels.

We attempt to cover most of the ways questions may be formulated on each topic, featuring those in which probabilities are explicitly provided from the start, and also questions posed in terms of 'enumeration of sets', by which we mean:

a a set of possible elements is provided, such as students in a class, and we assume one of these elements is going to be drawn at random

b working out the probability of a particular outcome, say that the randomly chosen student is female, boils down to counting the relevant elements that lead to that outcome, i.e. the number of female students.

We provide multiple methods of solution where appropriate, and include some notes identifying issues that students find difficult.

We provide a range of questions, covering the curriculum normally examined in public examinations. The topics covered are:

Chapter 8   probability scale: verbal and numerical

Chapter 9   event as 'favourable outcome' within an 'equally likely' sample space: simple enumeration of sets, more complex sample spaces, given ratios, 'inverse' problems★

Chapter 10   understanding fixed probability

Chapter 11   complement: enumeration of sets, probability

Chapter 12   mutually exclusive events adding to 1: enumeration of sets, probability, inverse problems

Chapter 13   multiple attributes – union and intersection: enumeration of sets, proportions, probability

Chapter 14   expectation

Chapter 15   estimating probabilities from experiments: relative frequency, larger experiments providing more accuracy

Chapter 16   identical independent events: sample space and enumeration, enumeration and multiplication, expected frequency trees, probability trees

Chapter 17 two non-identical independent events: sample space and enumeration of sets, probability, probability tree, inverse problems

Chapter 18 two dependent events: probability tree, frequency tree, inverse problems with different information provided

Chapter 19 inverse conditional probability – Bayes theorem: enumeration of sets, proportions and sets, frequency tree, expected frequency tree, probability tree.

★ Note that 'inverse' problems refer to questions in which the final outcome is provided and one of the inputs has to be obtained.

The probability scale

8.1 Verbal probability scale

Verbal expressions for probability are a good classroom discussion point, but terms such as 'likely', 'unlikely', 'possible', 'probable' and so on should not really be given any position on a probability scale, since their interpretation depends entirely on context (see Chapter 23: *How probable is probable?*). The only verbal terms that have an unambiguous location on a scale are 'impossible', 'evens' and 'certain'. Unfortunately, questions containing other terms do come up, in which case 'unlikely' can be taken as meaning low probabilities (say below $\frac{1}{4}$), and 'likely' can be taken as meaning high probabilities (say above $\frac{3}{4}$). It is reasonable to be asked 'which is more likely', as this means 'which has the greater probability'.

1   Look at the spinner in Figure 8.1.

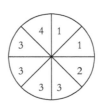

Figure 8.1

a   Circle the word that best describes the chance of spinning a '3'.

Impossible   Unlikely   Evens   Likely   Certain

b   Circle the word that best describes the chance of spinning a '4'.

Impossible   Unlikely   Evens   Likely   Certain

c   Circle the word that best describes the chance of spinning a '5'.

Impossible   Unlikely   Evens   Likely   Certain

d   Which is more likely, to spin a '1' or to spin a '4'?

Answer

1   a   Impossible   Unlikely   (Evens)   Likely   Certain

     b   Impossible   (Unlikely)   Evens   Likely   Certain

     c   (Impossible)   Unlikely   Evens   Likely   Certain

     d   It is more likely to spin a 1 than a 4.

8.2 Numerical probability scale

2   Lucas spins the spinner in Figure 8.2 once.

Figure 8.2

a   On the probability scale (shown in Figure 8.3), mark with a cross
✖ the probability of spinning a '1'.

b   On the probability scale, mark with a cross ✖ the probability of
spinning at least 3.

c   On the probability scale, mark with a cross ✖ the probability of
spinning an odd number.

Figure 8.3

Answer

2   a   The probability is marked on Figure 8.4.

Figure 8.4

b   The probability is marked on Figure 8.5.

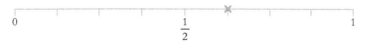

Figure 8.5

c   The probability is marked on Figure 8.6.

Figure 8.6

Event as 'favourable outcome' within an 'equally likely' sample space

9.1 Simple enumeration of sets

1 You have four 10p coins and three 20p coins in your pocket. You pick a coin at random. What is the probability it is a 20p coin?

Answer

1 You have 7 coins in total, 3 of which are 20p. So the probability of picking a 20p is $\frac{3}{7}$.

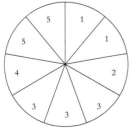

Figure 9.1

2 Consider the spinner shown in Figure 9.1.

Put a number in each of the spaces to make each of these statements true.

a There is a probability $\frac{1}{3}$ that the spinner will land on _____.

b There is probability 0 that the spinner will land on _____.

c The probability of landing on _____ is twice the probability of landing on _____.

Answer

2 a There are 9 possible outcomes for the spinner. 3 of them are *3*. So there is a probability $\frac{3}{9} = \frac{1}{3}$ that the spinner will land on *3*.

b There is probability 0 that the spinner will land on *any number other than 1, 2, 3, 4, 5.*

c *1* occurs twice as often as *2* on the spinner, so the probability of landing on *1* is twice the probability of landing on *2*. One answer is therefore the pair *(1,2)*; other possible answers are *(1,4)*, *(5,2)* and *(5,4)*.

3   Consider the following list of numbers.

   7   12   15   23   32   33   34   50

One of these numbers is picked at random. What is the probability that the picked number is:

a   odd?

b   greater than 6?

c   a prime number?

Answer

3   **a**   Out of the 8 numbers, 4 are odd. So the probability that the picked number is odd is $\frac{4}{8} = \frac{1}{2}$.

   b   Out of the 8 numbers, all are greater than 6. So the probability that the picked number is greater than 6 is $\frac{8}{8} = 1$.

   c   Out of the 8 numbers, 2 are prime (7 and 23). So the probability that the picked number is a prime number is $\frac{2}{8} = \frac{1}{4}$.

9.2 Complex sample space

4   Aisha, Benjamin, Charlotte and Daniel (A, B, C and D) form a club. A club chair and treasurer need to be chosen.

   a   Complete Table 9.1 to show the possible combinations of people to take on the roles.

Chair	Treasurer
A	B

Table 9.1

   b   If the chair and treasurer are chosen at random, what is the probability that Aisha or Benjamin get at least one of the posts?

4   **a**   The complete set of combinations is shown in Table 9.2.

Chair	Treasurer
A	B
A	C
A	D
B	A
B	C
B	D
C	A
C	B
C	D
D	A
D	B
D	C

Table 9.2

  b   By enumeration, 10 of these 12 combinations include Aisha or Benjamin. So the probability that Aisha or Benjamin get one of the posts is $\frac{10}{12} = \frac{5}{6}$. Alternatively, there are just two possibilities that do not include Aisha or Benjamin, so the probability that at least one of them is chosen is $1 - \frac{2}{12} = \frac{10}{12} = \frac{5}{6}$.

9.3 Information in the form of a ratio

5   There are four times as many girls as boys in the school choir. If I pick a choir member at random, what is the probability I pick a girl?

5   Frequency: suppose there are 10 boys. Then there are 40 girls in the choir, which then has a total size of 50. Therefore the probability of picking a girl is $\frac{40}{50} = 0.8$.

Ratio: the ratio of girls to boys is 4 to 1, so $\frac{4}{5}$ of the choir are girls and $\frac{1}{5}$ are boys. The probability of picking a girl is therefore $\frac{4}{5}$.

Algebraic: let the proportion of boys be x. Then the proportion of girls is $4x$. Together these proportions add to 1, so $x + 4x = 1$, and hence $x = 0.2$. This means the probability of picking a girl is 0.8.

9.4 Inverse problems – enumeration of sets from given probabilities

6 A bag of 50 sweets contains a mixture of red and blue sweets. If I take a sweet at random, the probability of getting a red one is 0.4. How many blue sweets are in the bag?

Answer

6 The proportion of red sweets in the bag is 0.4, and so there are $0.4 \times 50 = 20$ red sweets. Therefore there are 30 blue sweets.

Alternatively, if $\frac{4}{10}$ are red, then $\frac{6}{10}$ are blue. If there are 50 sweets, then $\frac{1}{10} = 5$ sweets so $\frac{6}{10} = 30$ sweets.

7 The spinner in Figure 9.2 has 6 equal sections. To complete the spinner you must write one number in each section. You must use three different numbers. The probability that the spinner lands on an even number is less than the probability that it lands on an odd number. The most likely number to land on is a 3. Complete the spinner to show one possible arrangement of the numbers.

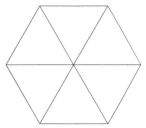

Figure 9.2

Answer

7 One possible solution is shown in Figure 9.3.

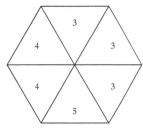

Figure 9.3

Alternatively, there could be four 3s and two different single numbers, which could be odd or even.

Chapter 1

Understanding fixed probability

1   Amelia takes a **fair** coin and flips it 4 times. It comes up

Heads   Tails   Tails   Tails

What is the probability that the next time the coin is flipped it will come up Tails?

Answer

1   The probability that it will be Tails is $\frac{1}{2}$ as it is a fair coin. The previous results are irrelevant. Note: even though the term 'fair' is in bold, students often will say $\frac{3}{4}$, reflecting the relative frequency in the previous results.

2   Jack has thrown a **fair** die 12 times and a 'six' has not come up. He says 'a six must be due soon'. Is he right?

Answer

2   Jack is wrong. The probability of a 'six' coming up is $\frac{1}{6}$, regardless of what has happened before.

Complement

11.1 Enumeration of sets

1   There are 30 students in a class that has been misbehaving, of whom
    18 are boys. If a student is picked at random from the class to be sent
    to the headteacher, what is the probability it is a girl?

Answer

1   There are $30 - 18 = 12$ girls in the class. Therefore the probability
    that a randomly selected student is a girl is $\frac{12}{30} = \frac{2}{5}$.

11.2 Probability

2   If you throw a drawing-pin in the air, the probability of it landing
    with its point up is 0.35. What is the probability of it coming down
    without its point up?

Answer

2   By the complement rule, the probability of it coming down without
    its point up is $1 - 0.35 = 0.65$.

Chapter 12

Mutually exclusive events adding to 1 (more than two categories)

12.1 Simple enumeration of sets

1   Yusuf keeps a drawer full of loose, single socks. He has 5 white socks, 3 black socks, 2 green socks and 4 pink socks. He reaches in the drawer and takes one out at random, so that each sock has the same probability of being picked. What is the probability that he picks:

   a   a pink sock?

   b   a yellow sock?

Answer

1   a   There are $5 + 3 + 2 + 4 = 14$ socks in the drawer. 4 are pink. So the probability he picks a pink sock is $\frac{4}{14} = \frac{2}{7}$.

   b   There are no yellow socks, so the probability he picks one is 0.

12.2 All but one probability given

2   Reina has four types of music on her music player: rock, world, electronic and classical. When the player picks a track at random, the probabilities of picking the first three types are as follows.

Type of music	Rock	World	Electronic	Classical
Probability	0.45	0.20	0.30	

What is the probability that the player picks:

a   a classical track?

b   a rock or a classical track?

Answer

2 **a** These are mutually exclusive categories, so the probabilities must add to 1. This means the probability of a classical track is $1 - (0.45 + 0.20 + 0.30) = 1 - 0.95 = 0.05$.

   b The probability of picking a rock or a classical track is the sum of their probabilities: $0.45 + 0.05 = 0.50$.

12.3 Inverse problems: given a relationship, solve for probabilities

3 A tube of small chocolate sweets has four different colours.

- There are twice as many blue sweets as red sweets.
- There are three times as many green sweets as blue sweets.
- There are also some brown sweets.

The probability of choosing a red sweet is 0.05. What is the probability of picking a brown sweet?

Answer

3 The probability of picking a red sweet is 0.05.

There are twice as many blue as red, and so the probability of picking a blue sweet is 0.10.

There are three times as many green as blue, and so the probability of picking a green sweet is 0.30.

So the probability of picking a brown sweet is $1 - (0.05 + 0.10 + 0.30) = 0.55$.

This could also be answered by assuming there were 100 sweets in the tube. Of the 100, 5 will be red and the remaining quantities follow from the information given.

Colour of sweet	Red	Blue	Green	Brown
Proportion	x	$2x$	$6x$	
Number out of 100 sweets	5	10	30	55
Probability	0.05	0.10	0.30	0.55

12.4 Solving using formulae

We include, for completeness, the type of question that we hope would not be set as part of a modern assessment system (because this is simply an abstract exercise, lacking any context, and so students can have no idea if their answer makes sense or not).

4   A and B are mutually exclusive events, with the following properties:

[1] P(B) = 2 P(A)

[2] P(A or B) = 0.57

Find P(B).

Answer

4   Since A and B are mutually exclusive, P(A or B) = P(A) + P(B).
Hence, from [2], P(A) + P(B) = 0.57.
Since, from [1], P(B) = 2 P(A), we can see that 3P(A) = 0.57.
So P(A) = 0.19 and P(B) = 0.57 − 0.19 = 0.38.

Multiple attributes – union and intersection

13.1 Enumeration of sets

1   Paige has recorded 30 matches of her favourite football team, Milltown Rovers. In 8 matches Milltown did not score, but their opponents did. In 10 matches Milltown scored, but their opponents did not. In 5 matches neither team scored.

   a   Use this information to complete the Venn diagram in Figure 13.1.

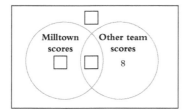

Figure 13.1

   In how many of these matches did both teams score?

   b   If Paige chooses a match at random to watch again, what is the probability that she picks a match in which her team scores?

Answer

1   a   The Venn diagram can be completed using each piece of information in turn, including the information that there were 30 matches altogether (see Figure 13.2).

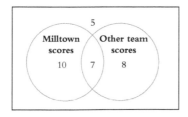

Figure 13.2

   We see from the Venn diagram that there are 7 matches in which both teams score.

   b   There are $10 + 7 = 17$ matches in which Milltown scores. So if Paige chooses a match at random to watch again, the probability that she picks a match in which her team scores is $\frac{17}{30}$.

2   A year group of 80 students is asked whether they are left-handed or right-handed, and whether they prefer maths or English. Table 13.1 summarises how they responded.

	Left-handed	Right-handed	Total
Prefer maths	6		45
Prefer English			
Total		65	80

Table 13.1

a   Complete Table 13.1.

b   If I pick a student at random, what is the probability that they are left-handed?

c   If I pick student at random, what is the probability they are right-handed **and** prefer English?

Answer

2   a   Table 13.2 shows the complete set of responses.

	Left-handed	Right-handed	Total
Prefer maths	6	39	45
Prefer English	9	26	35
Total	15	65	80

Table 13.2

b   There are 15 left-handed students out of 80, so if I pick a student at random, the probability that they are left-handed is $\frac{15}{80} = \frac{3}{16}$.

c   There are 26 students who are right-handed **and** prefer English, so the probability of picking such a student is $\frac{26}{80} = \frac{13}{40}$.

3   40 students are asked whether they watch football, tennis or rugby on TV.

· 10 say they don't watch any sport on TV.

· 5 watch all three sports.

· 15 watch football and tennis, but do not watch rugby.

· 6 watch football and rugby.

· 24 watch football.

· 7 watch tennis and rugby.

· 4 watch only tennis.

a   Complete the Venn diagram in Figure 13.3.

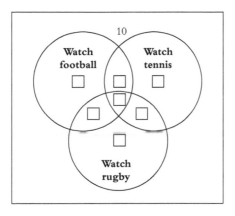

Figure 13.3

How many students watch only rugby?

b   If I pick a student at random, what is the probability they watch tennis?

Answer

3   a   Students need to be clear that, for instance, '6 watch football and rugby' includes both those who also watch tennis and those who do not. Working steadily through the items of information gives the Venn diagram in Figure 13.4.

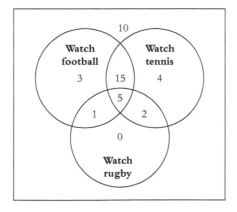

Figure 13.4

By subtracting from the total of 40, we find that no students watch only rugby.

b   There are 4 + 15 + 5 + 2 = 26 students who watch tennis. So if I pick a student at random, the probability they watch tennis is $\frac{26}{40} = \frac{13}{20} = 0.65$.

13.2 Proportions given

4 In a school:

- 20% of the students are boys who walk to school
- 45% of the students are boys
- 60% of students walk to school.

a Display this information on the Venn diagram in Figure 13.5.

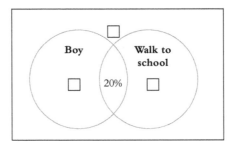

Figure 13.5

b If I pick a student at random from the school, what is the probability that this is a girl who does not walk to school?

Answer

4 a The Venn diagram is shown in Figure 13.6. The first item of information, 20%, gives the intersection of 'boy' and 'walking'. The other items give the percentage who are boys who do not walk to school, and the percentage who are girls who walk to school.

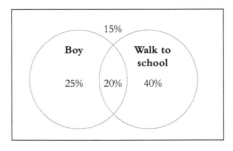

Figure 13.6

Note: a standard mistake is to draw the two circles as representing boys and girls, which of course cannot have an intersection. But in this case the Venn diagram is given.

b 15% of students are girls who do not walk to school. So if I pick a student at random from the school, the probability that she is a girl who does not walk to school is 15% = 0.15.

13.3 Given probabilities

5   A large tray contains sandwiches. Each sandwich has one of three
     sorts of bread, and one of three kinds of filling. Maria closes her eyes
     and picks a sandwich at random. The probability that she gets each
     type of sandwich is shown in Table 13.3.

	Egg	Tomato	Cheese
White bread	0.04	0.08	0.12
Brown bread	0.14	0.20	0.15
Granary bread	0.06	0.10	

Table 13.3

  a   What is the probability that Maria takes an egg sandwich?

  b   What is the probability that Maria takes a sandwich made of
      white or brown bread?

  c   What is the probability that Maria takes a cheese and granary
      bread sandwich?

Answer

5   a   The probability that Maria takes an egg sandwich is
        $0.04 + 0.14 + 0.06 = 0.24$.

    b   The probability that Maria takes a sandwich made of white bread
        is $0.04 + 0.08 + 0.12 = 0.24$. The probability that Maria takes a
        sandwich made of brown bread is $0.14 + 0.20 + 0.15 = 0.49$. So
        the total probability of taking either a white or a brown bread
        sandwich is $0.24 + 0.49 = 0.73$.

    c   The probabilities must add to 1, and so the probability
        that Maria takes a cheese and granary bread sandwich is
        $1 - (0.73 + 0.06 + 0.10) = 0.11$.

Alternatively, students could complete the table first, then read the
answers off it.

Chapter 14

Expectation

1   Olivier flips a fair coin 200 times.

   a   How many times does he expect to get a Head?

   b   Olivier gets 110 Heads and claims the coin is not fair because it has a bias towards getting Heads. Is this reasonable?

Answer

1   a   Since the coin is fair, the probability of getting a Head is $\frac{1}{2}$. So in 200 flips, Olivier expects half to be Heads, that is, 100 Heads.

   b   Olivier gets 110 Heads, but this does not mean the coin is not fair. He should not be surprised not to get exactly 100 Heads (but if he got 150, then something would be odd).

2   a   Hanna is going to throw a fair die 36 times. How many times does she expect to throw a 'six'?

   b   Sara is going to throw a pair of fair dice 36 times. How many times does she expect to throw a 'double-six'?

   c   If, after 36 throws of a pair of dice, Sara has not thrown a 'double-six', what is the probability of her throwing a 'double-six' at the next throw?

Answer

2   a   The probability of throwing a 'six' at a single attempt is $\frac{1}{6}$. So over the 36 attempts Hanna would expect to throw $36 \times \frac{1}{6} = 6$ 'sixes'.

   b   The probability of throwing a 'double-six' at a single attempt is $\frac{1}{6} \times \frac{1}{6} = \frac{1}{36}$. So over the 36 attempts Sara would expect to throw $36 \times \frac{1}{36} = 1$ 'double-six'.

   c   If, after 36 throws of a pair of dice, Sara has not thrown a 'double-six', the probability that she throws a 'double-six' at the next throw is still $\frac{1}{36}$.

Estimating probabilities from experiments

15.1 Relative frequency

1 Jon starts to note the colours of cars going down the road. He planned to count 100 cars, but became bored after 45. Table 15.1 shows his results.

Colour of car	White	Silver	Red	Any other colour
Number of cars	17	12	6	10

Table 15.1

   a Estimate the probability that the next car is silver.

   b How could this estimate be made more reliable?

Answer

1 a Jon counted 45 cars, of which 12 were silver. So an estimate of the probability that the next car is silver is $\frac{12}{45} = \frac{4}{15}$.

   b This estimate can be made more reliable by counting more cars.

15.2 Larger experiments provide more accuracy

2   The spinner in Figure 15.1 has white, black and grey regions. When Grace and Ethan spin it, they get the results shown in Table 15.2.

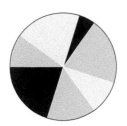

Figure 15.1

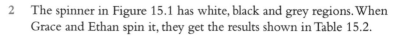
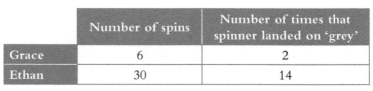

	Number of spins	Number of times that spinner landed on 'grey'
Grace	6	2
Ethan	30	14

Table 15.2

They want to estimate the probability of the spinner landing on 'grey'.

a   Grace says: There are three possible outcomes, and so the probability of landing on "grey" is $\frac{1}{3}$. And my results prove it.' What is wrong with Grace's opinion?

b   Ethan says 'the probability of landing on "grey" is $\frac{14}{30}$'. Why is this a more reliable estimate than Grace's?

c   What might be a better estimate than $\frac{14}{30}$?

Answer

2   a   Just because there are 3 possible outcomes does not mean they are equally likely – clearly the grey area is greater than either the black or the white area on the spinner. Grace's 2 out of 6 landing on 'grey' is based on only 6 spins and so is not reliable, and it does not prove the probability is $\frac{1}{3}$.

b   Ethan's estimate is more reliable as it is based on more spins than Grace's opinion.

c   A better estimate could be obtained by pooling the data from Grace and Ethan to get an estimate of $\frac{16}{36} = \frac{4}{9}$.

Identical independent events

16.1 Sample space and enumeration

1 Faiza throws a fair die and writes down the number that comes up.
   She throws it again and writes down this second number. She then
   subtracts the second number from the first number.

   a Complete Table 16.1 showing the possible outcomes.

		Number from first throw					
		1	2	3	4	5	6
Number from second throw	1	0	1				
	2	−1					
	3						
	4						
	5						
	6						

Table 16.1

   b What is the probability that her total after subtraction equals 0?

   c What is the probability that her total after subtraction is greater
      than 4?

   d What is the probability that her total after subtraction is less than −3?

Answer

1 a The complete set of possible outcomes is shown in Table 16.2.

		Number from first throw					
		1	2	3	4	5	6
Number from second throw	1	0	1	2	3	4	5
	2	−1	0	1	2	3	4
	3	−2	−1	0	1	2	3
	4	−3	−2	−1	0	1	2
	5	−4	−3	−2	−1	0	1
	6	−5	−4	−3	−2	−1	0

Table 16.2

b Out of 36 possible outcomes, six are 0, so the probability that her total after subtraction equals 0 is $\frac{6}{36} = \frac{1}{6}$.

c Out of 36 possible outcomes, one is greater than 4, so the probability that her total after subtraction is greater than 4 is $\frac{1}{36}$.

d Out of 36 possible outcomes, three are less than -3, and so the probability that her total after subtraction is less than -3 is $\frac{3}{36} = \frac{1}{12}$.

Note that some students may mistakenly include '4' in 'greater than 4'.

16.2 Enumeration and multiplication

2   Ruby has a bag of 20 sweets: 6 toffees and 14 chocolates. She invites Tariq to pick a sweet at random, so each sweet has the same probability of being chosen. Tariq doesn't like toffees: if he picks a toffee he will put it back in the bag and pick a sweet at random again. What is the probability that Tariq picks a toffee and then, after putting it back, picks a toffee again?

Answer

2   The probability Tariq picks a toffee the first time is $\frac{6}{20} = \frac{3}{10}$. After he has put it back, the probability that he picks a toffee at the second attempt is also $\frac{3}{10}$. So, by the multiplication rule, the probability that he picks toffees both times is $\frac{3}{10} \times \frac{3}{10} = \frac{9}{100}$. Alternatively, look at the expected outcomes over 100 repetitions of the same situation.

16.3 Expected frequency tree

3   20 students in a class are going to flip a fair coin twice.

   a   Complete the expected frequency tree shown in Figure 16.1.

Figure 16.1

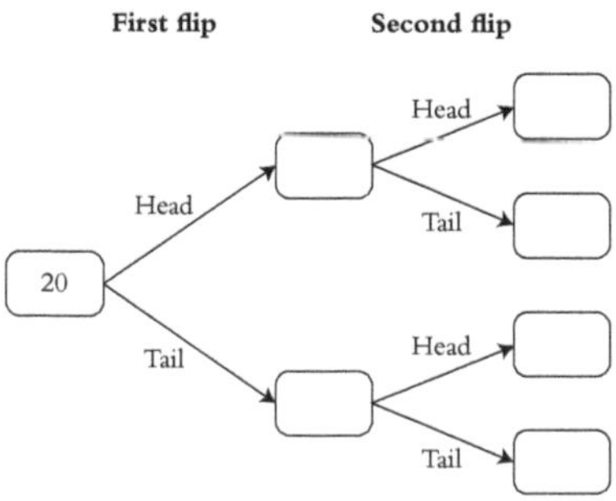

   b   How many students should we expect to get at a Head on both flips?

   c   How many students should we expect to get at least one Head?

Answer

3 a The expected frequency tree is shown in Figure 16.2

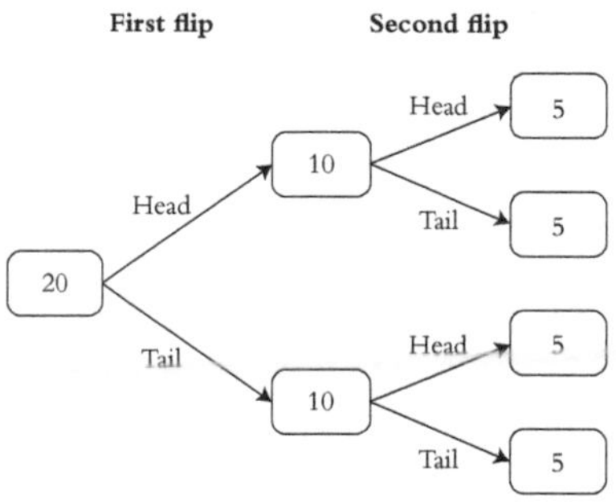

Figure 16.2

First flip   Second flip

Head → 5
10
Head
Tail → 5
20
Head → 5
Tail
10
Tail → 5

b We expect 5 students to get at a Head on both flips.

c We expect 15 students to get at least one Head.

16.4 Probability tree

4 The probability that it rains in the morning is 0.2, and there is the same probability of rain in the afternoon, independent of what happened in the morning.

a Complete the probability tree shown in Figure 16.3.

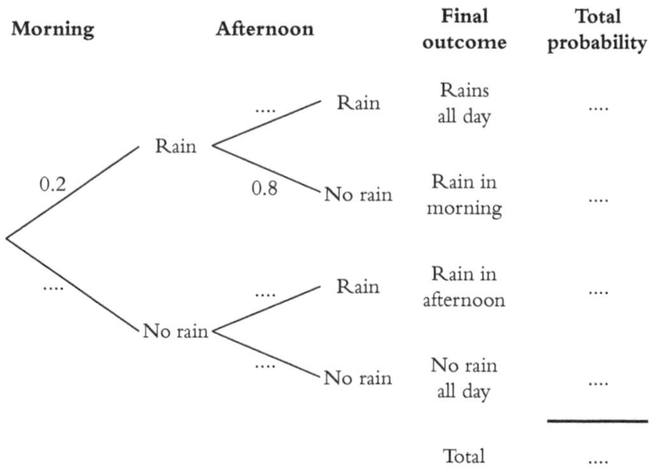

Figure 16.3

Morning	Afternoon		Final outcome	Total probability
	.... Rain		Rains all day	....
Rain				
0.2	0.8 No rain		Rain in morning	....
	.... Rain		Rain in afternoon	....
.... No rain				
	.... No rain		No rain all day	....
			Total	....

b What is the probability that it rains at some point in the day?

4 a The completed probability tree is shown in Figure 16.4.

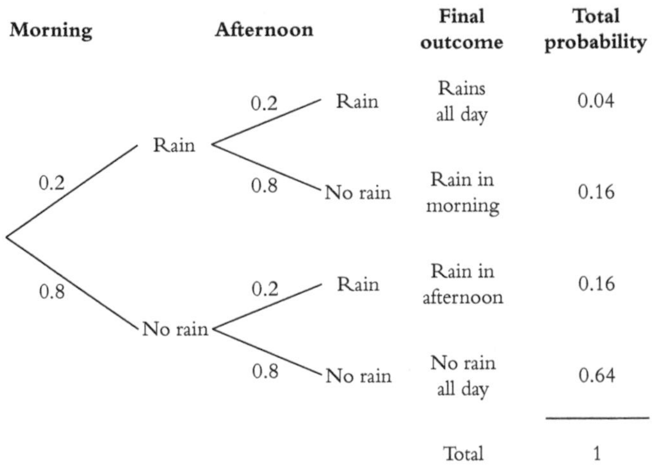

Morning	Afternoon	Final outcome	Total probability
	0.2 — Rain	Rains all day	0.04
0.2 — Rain <	0.8 — No rain	Rain in morning	0.16
0.8 — No rain <	0.2 — Rain	Rain in afternoon	0.16
	0.8 — No rain	No rain all day	0.64
		Total	1

Figure 16.4

b The probability that it rains at some point in the day is $0.04 + 0.16 + 0.16 = 0.36$. Alternatively, the probability that it does not rain at some point in the day is 0.64, so the probability that it does is $1 - 0.64 = 0.36$ by the complement rule.

5 Dullchester United players Jose, Dean and Wayne are competing in a penalty shootout based on three attempts. Each, independently, has probability 0.8 of scoring. The other team score in only one of their three penalties.

a Complete the probability tree (in Figure 16.5) for the Dullchester United players.

Jose's shot	Dean's shot	Wayne's shot	Final outcome	Total probability
		0.8 — Score	3 goals	0.512
0.8 — Score	0.8 — Score			
0.2 — Miss				

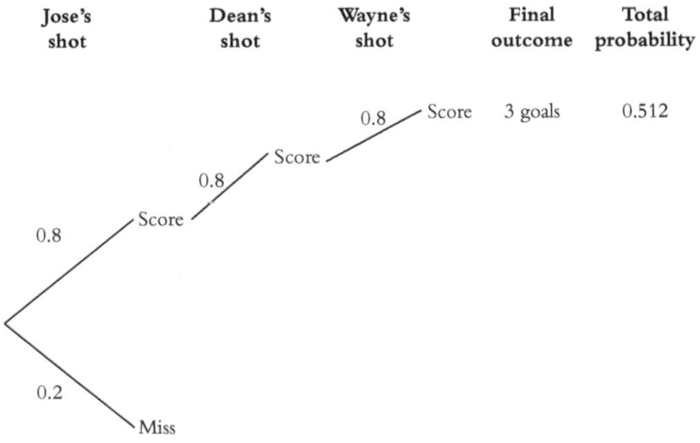

Figure 16.5

b What is the probability that Dullchester score more than once and so win the shootout?

Answer

5 a Figure 16.6 shows a completed probability tree of the three events.

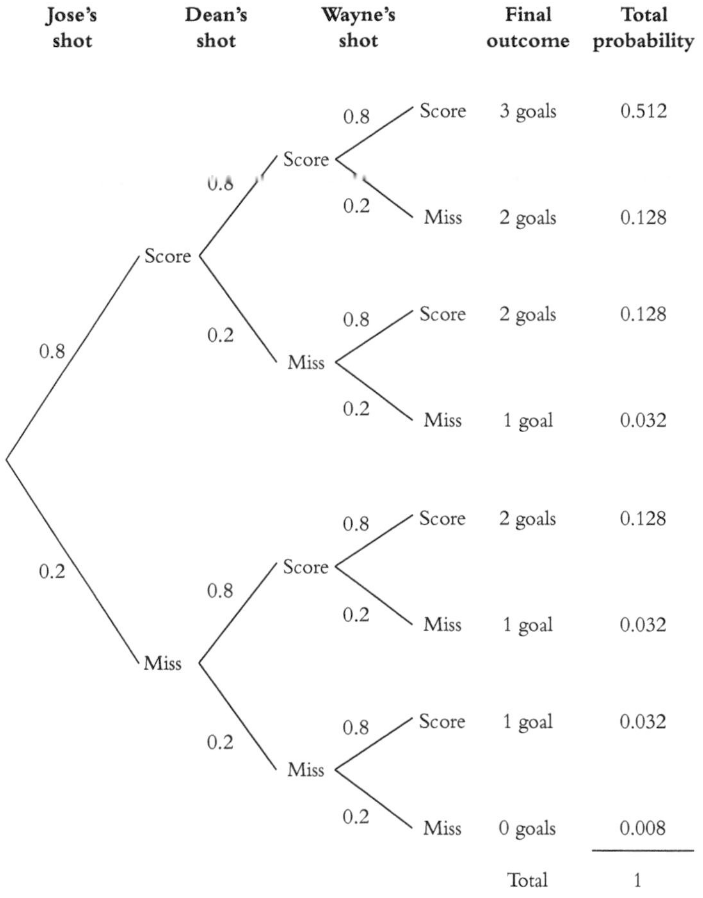

Jose's shot	Dean's shot	Wayne's shot	Final outcome	Total probability
		0.8 Score	3 goals	0.512
	0.8 Score	0.2 Miss	2 goals	0.128
0.8 Score	0.2 Miss	0.8 Score	2 goals	0.128
		0.2 Miss	1 goal	0.032
	0.8 Score	0.8 Score	2 goals	0.128
0.2 Miss		0.2 Miss	1 goal	0.032
	0.2 Miss	0.8 Score	1 goal	0.032
		0.2 Miss	0 goals	0.008
			Total	1

Figure 16.6

b Dullchester need to score at least twice. The total probability of scoring at least 2 goals is

$0.512 + 0.128 + 0.128 + 0.128 = 0.896.$

Alternatively, we could reason without a tree that the four outcomes that would lead to at least two of them scoring are SSS, MSS, SMS and SSM, where S = score and M = miss. The probability of them all scoring (SSS) is $0.8 \times 0.8 \times 0.8 = 0.512$.

The probability of MSS = $0.2 \times 0.8 \times 0.8 = 0.128$, and this is also the probability of SMS and SSM. So the total probability is $0.512 + 0.128 + 0.128 + 0.128 = 0.896$.

Two non-identical independent events

10.1 Establishing a sample space and enumeration of sets

1   Isabella has the two fair spinners as shown in Figure 17.1.

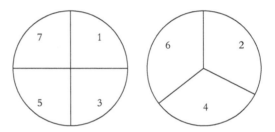

Figure 17.1

She spins each once and adds the results to give a total score.
Complete Table 17.1 to show all possible scores.

		Number on Spinner 2		
		2	**4**	**6**
	1			
Number on Spinner 1	**3**			
	5	7		
	7			
			Total score	

Table 17.1

What is the probability that Isabella scores:

a   at least 7?

b   an odd number?

Answer

1   The scores are shown in completed Table 17.2.

		Number on Spinner 2		
		2	4	6
	1	3	5	7
Number on Spinner 1	3	5	7	9
	5	7	9	11
	7	9	11	13
		Total score		

Table 17.2

a   There are 12 equally likely possible outcomes, of which 9 are at least 7. So the probability of scoring at least 7 is $\frac{9}{12} = \frac{3}{4}$. Alternatively, we could think of what we expect to happen in 12 plays of the game: we would expect to score at least 7 in 9 of these games.

b   All 12 outcomes are odd, so the probability of getting an odd-numbered score is $\frac{12}{12} = 1$.

2   Nadiyah has three coins in her pocket: 5p, 10p and 50p. Alexander also has three coins in his pocket, one 10p and two 20p coins. A packet of sweets is 30p. If each of them takes a single coin at random from their pocket, what is the probability that together they will have enough to buy a packet of sweets?

Answer

2   A table of possible outcomes (Table 17.3) has to be constructed, taking into account that Alexander has two coins of the same value.

		Alexander's coin		
		10p	20p	20p
	5p	15p	25p	25p
Nadiyah's coin	10p	20p	30p	30p
	50p	60p	70p	70p
		Total amount		

Table 17.3

There are 9 equally likely outcomes, of which 5 lead to them having enough to buy the sweets. The probability is therefore $\frac{5}{9}$.

3 A game offers a prize if when you throw a die you get a six *and* when you flip a coin you get a Head. The probability tree in Figure 17.2 shows the possible outcomes when you play.

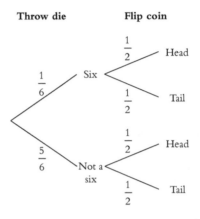

Figure 17.2

a Complete the tree to show the probabilities of all four outcomes.

b What is the probability that you win, that is, throw a six and flip a Head?

Answer

3 a The completed tree is shown in Figure 17.3.

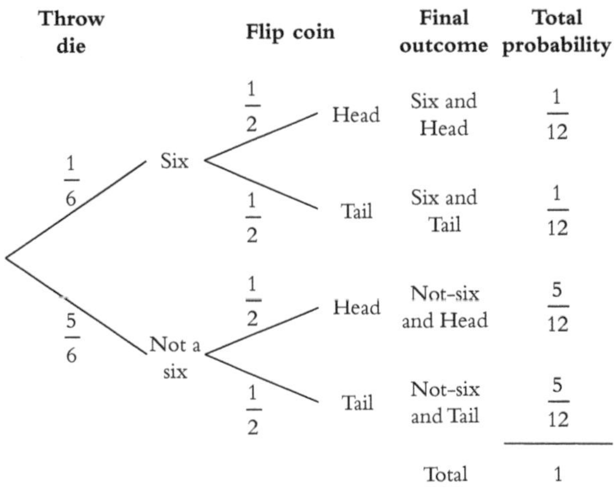

Throw die	Flip coin	Final outcome	Total probability
Six ($\frac{1}{6}$)	Head ($\frac{1}{2}$)	Six and Head	$\frac{1}{12}$
	Tail ($\frac{1}{2}$)	Six and Tail	$\frac{1}{12}$
Not a six ($\frac{5}{6}$)	Head ($\frac{1}{2}$)	Not-six and Head	$\frac{5}{12}$
	Tail ($\frac{1}{2}$)	Not-six and Tail	$\frac{5}{12}$
		Total	1

Figure 17.3

b The probability of getting a six and a Head is found by multiplying along the top branch, so $\frac{1}{6} \times \frac{1}{2} = \frac{1}{12}$.

4   Mona is going on a gap year and buys travel insurance for theft
    and illness. The probability that she will be robbed is 0.1, and the
    probability that she falls ill is 0.2. These events are independent of
    each other.

    a   What is the probability she will make any claim on her
        insurance?

    b   Out of 100 similar travellers, how many would you expect to
        make claims for **both** being robbed and falling ill?

Answer

4   a   The only way Mona does not make a claim is if she is not
        robbed **and** does not fall ill, which happens with probability
        $0.9 \times 0.8 = 0.72$, since the events are independent. Therefore the
        probability that she claims is $1 - 0.72 = 0.28$.

        Alternatively, we could complete the tree, as in Figure 17.4,
        giving a $0.02 + 0.08 + 0.18 = 0.28$ probability of claiming.

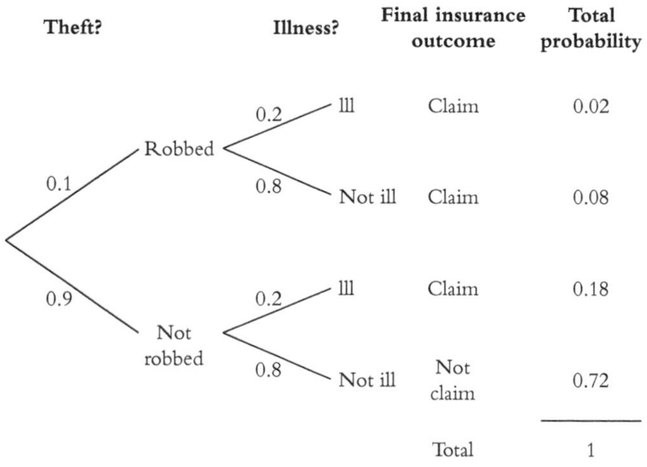

Theft?	Illness?	Final insurance outcome	Total probability
	0.2 — Ill	Claim	0.02
Robbed			
0.1	0.8 — Not ill	Claim	0.08
	0.2 — Ill	Claim	0.18
0.9 — Not robbed			
	0.8 — Not ill	Not claim	0.72
		Total	1

Figure 17.4

    b   From the tree, there is 0.02 probability of making both claims.
        So out of 100 similar travellers, we would expect 2 to claim for
        both theft and illness.

Alternatively, we could answer the whole question by considering what we would expect to happen to 100 similar travellers, using the expected frequency tree in Figure 17.5.

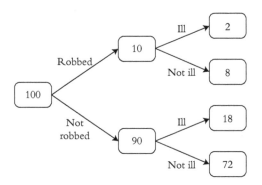

Figure 17.5

17.3 Inverse: Probability tree, given probabilities of final outcomes

5   Applicants for a job have to pass two tough test papers. 70% of people fail the first paper, and 42% of people fail both papers. I pick an applicant at random. By completing the tree in Figure 17.6, and assuming the test results are independent, find the probability that the applicant gets a job.

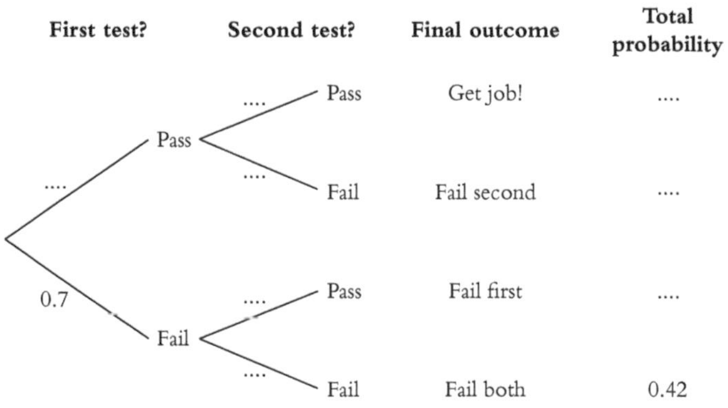

First test?	Second test?	Final outcome	Total probability
	.... Pass	Get job!	....
Pass			
	.... Fail	Fail second	....
0.7	.... Pass	Fail first	....
Fail			
	.... Fail	Fail both	0.42

Figure 17.6

Answer

5 First, the probability of passing the first test is 0.3, by the complement rule. Second, the probability of failing the second test is $0.42 \div 0.70 = 0.6$, since the probabilities along the bottom branch must multiply to 0.42. This means the probability of passing the second test is 0.4 (by the complement rule). Since the tests are independent, the probability of passing the second test does not depend on the result of the first test, and so we can complete the tree (see Figure 17.7). Then the overall probability of passing both tests is $0.3 \times 0.4 = 0.12$.

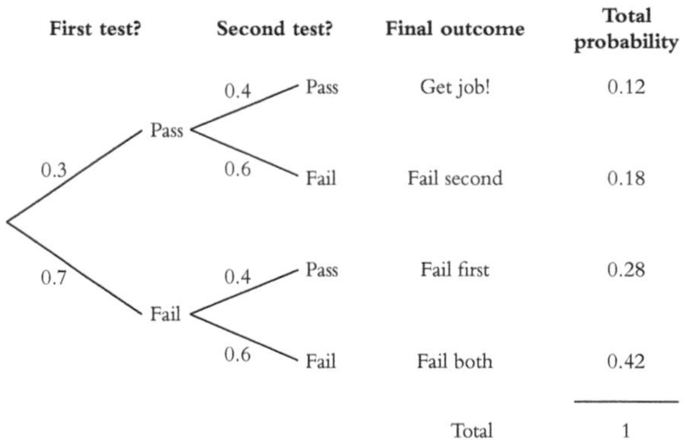

First test?
Second test?
Final outcome
Total probability
0.3
Pass
0.4
Pass
Get job!
0.12
0.6
Fail
Fail second
0.18
0.7
Fail
0.4
Pass
Fail first
0.28
0.6
Fail
Fail both
0.42
Total
1

Figure 17.7

So the probability of a random applicant getting a job is 0.12.

We could also solve this by examining what we would expect to happen to 100 people using the frequency tree in Figure 17.8.

100
Pass first test
30
Pass second test
12
Fail second test
18
Fail first test
70
Pass second test
28
Fail second test
42

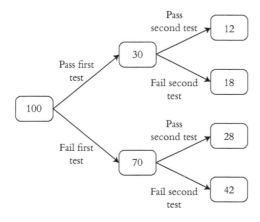

Figure 17.8

17.4 Solving from formulae alone

We return to the difficult question featured in Chapter 1.

6   Consider three events A, B and C. A and B are independent, B and C are independent and A and C are mutually exclusive. Their probabilities are P(A) = 0.3, P(B) = 0.4 and P(C) = 0.2. Calculate the probability of the following events.

     a   Both A and C occur.

     b   Both B and C occur.

     c   At least one of A or B occur.

Answer

6   **a**   P(A and C) = 0, since A and C are mutually exclusive.

     b   P(B and C) = P(B) × P(C) since B and C are independent. And so P(B and C) = 0.4 × 0.2 = 0.08.

     c   P(at least one of A or B) = 1 − P(neither A nor B occur) by complement rule.

         P(neither A nor B occur) = P(not A) × P(not B) since A and B are independent.

         So P(at least one of A or B) = 1 − P(not A) × P(not B) = 1 − (0.7 × 0.6) = 1 − 0.42 = 0.58.

We hope your students are never asked this sort of question.

Chapter 18

Two dependent events

18.1 Fully specified probability tree

1   The probability that it rains in the morning is 0.3. If it rains in the morning, the probability that it carries on raining in the afternoon is 0.5. If it does not rain in the morning, the probability that it carries on being fine in the afternoon is 0.9.

a   Complete the probability tree in Figure 18.1 using this information.

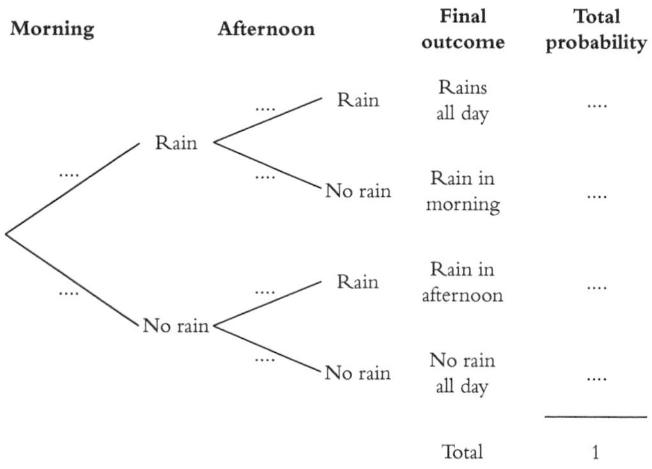

Morning	Afternoon	Final outcome	Total probability
	Rain .... → Rain	Rains all day	....
Rain			
	.... → No rain	Rain in morning	....
	.... → Rain	Rain in afternoon	....
No rain			
	.... → No rain	No rain all day	....
		Total	1

Figure 18.1

b   What is the probability that the weather is the same in the afternoon as in the morning?

c   What is the probability that it rained at some time in the day?

d   If it rained during the day, what is the probability that it rained in the morning?

Answer

1 **a** The completed probability tree is shown in Figure 18.2.

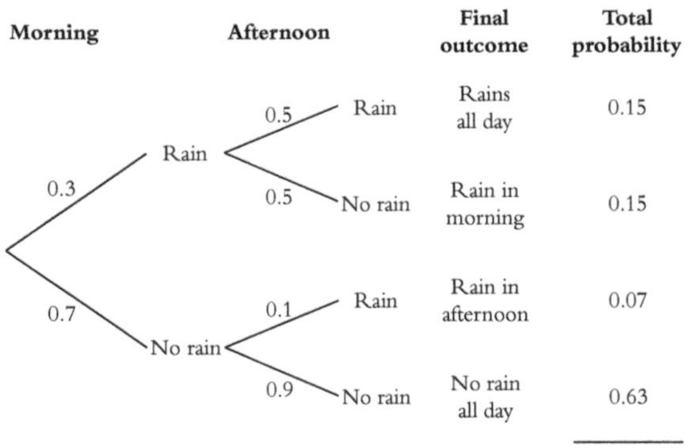

Morning
Afternoon
Final outcome
Total probability
0.5
Rain
Rains all day
0.15
Rain
0.3
0.5
No rain
Rain in morning
0.15
0.1
Rain
Rain in afternoon
0.07
0.7
No rain
0.9
No rain
No rain all day
0.63
Total
1

Figure 18.2

b The probability that the weather is the same in the afternoon as in the morning is $0.15 + 0.63 = 0.78$ (adding over mutually exclusive events).

c The probability that it rained at some time in the day is $0.15 + 0.15 + 0.07 = 0.37$, or use the complement, that $1 - 0.63 = 0.37$.

d On 37% of days it rained during the day, and on 30% it rained in the morning. So the probability that it rained in the morning, given that it rained at some time, is $\frac{0.30}{0.37} = 0.81$ (2dp).

Alternatively, we can use an expected frequency tree based on what we would expect over 100 days. This is shown in Figure 18.3.

Morning
Afternoon
Rain
15
30
Rain
No rain
15
100
Rain
7
No rain
70
No rain
63

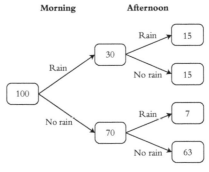

Figure 18.3

a The probability that the weather is the same in the afternoon as in the morning is $\frac{15 + 63}{100} = 0.78$.

b The probability that it rained at some time in the day is $\frac{15 + 15 + 7}{100} = 0.37$.

c We expect 37 days on which it rains during the day, and 30 when it rains in the morning. So the probability that it rained in the morning, given it rained at some time, is $\frac{30}{37}$.

2 Adam has 3 white single socks and 4 black single socks that he keeps loose in a drawer. If he takes out two socks at random in the morning, what is the probability he picks a pair of socks of the same colour?

Answer

2 The probability tree is shown in Figure 18.4.

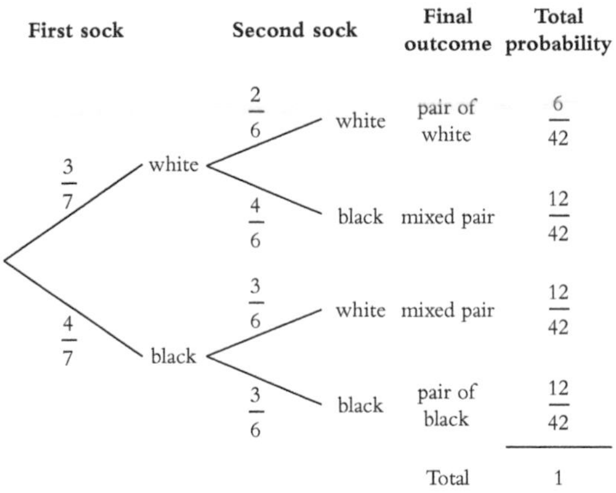

First sock	Second sock	Final outcome	Total probability

$\frac{3}{7}$ white

$\frac{2}{6}$ white — pair of white — $\frac{6}{42}$

$\frac{4}{6}$ black — mixed pair — $\frac{12}{42}$

$\frac{4}{7}$ black

$\frac{3}{6}$ white — mixed pair — $\frac{12}{42}$

$\frac{3}{6}$ black — pair of black — $\frac{12}{42}$

Total 1

Figure 18.4

The probability of getting a matching pair is therefore $\frac{6}{42} + \frac{12}{42} = \frac{1}{7} + \frac{2}{7} = \frac{3}{7}$. Note: common errors are to assume that the probabilities do not change after taking the first sock, and to change the numerator for each probability but not the denominator.

Alternatively, this could be solved using an expected frequency tree representing what we would expect to happen over 7 days (a week). This is shown in Figure 18.5.

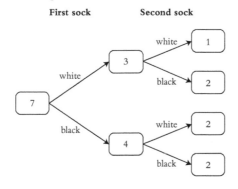

Figure 18.5

The answer follows immediately from this representation, but it requires more thought to realise, for example, that on three occasions of drawing a white sock first, we would expect to draw a further white sock on one of those occasions, as the probability is $\frac{2}{6} = \frac{1}{3}$ once the first sock has been removed.

3   A class of size 20 includes two sets of twins. The teacher picks two students at random to be class representatives. What is the probability that a matching set of twins is picked?

Answer

3   The probability that the first student picked is a twin is $\frac{4}{20} = \frac{1}{5}$. If this happens, there are now 19 students left, of which one is the other twin. So the probability that the second random choice is the matching twin is $\frac{1}{19}$. Therefore the overall probability of picking a pair of twins is $\frac{1}{5} \times \frac{1}{19} = \frac{1}{95}$.

This could be answered using a probability tree, but only the top branch needs to be considered (see Figure 18.6), giving the overall probability as $\frac{4}{380} = \frac{1}{95}$.

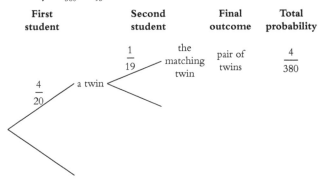

Figure 18.6

4   A bag has 10 counters numbered 1 to 10. Hussain picks two out at random, writes down their numbers and puts them back in the bag. Maryam then picks two counters out of the bag. What is the probability that Maryam picks the same counters as Hussain?

Answer

4   It is crucial that it does not matter what counters Hussain picked, so we can call them m and n. The probability that Maryam's first counter is m or n is $\frac{2}{10}$. If this happens, there are 9 counters left of which one is Hussain's remaining counter. So, given that the first was one of Hussain's counters, the probability that the second of Maryam's counters is the other one of Hussain's is $\frac{1}{9}$. So overall the probability of picking both of Hussain's counters is $\frac{2}{10} \times \frac{1}{9} = \frac{2}{90} = \frac{1}{45}$.

We can also use a probability tree, of which only the top branch is required (see Figure 18.7).

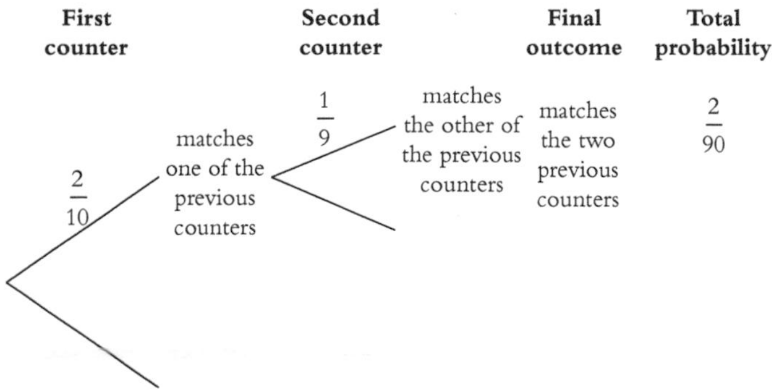

First counter	Second counter	Final outcome	Total probability

$\frac{2}{10}$ — matches one of the previous counters — $\frac{1}{9}$ matches the other of the previous counters — matches the two previous counters — $\frac{2}{90}$

Figure 18.7

Another method is to consider the number of ways to pick two counters, which is $10 \times 9 = 90$. So the probability of picking a particular pair is $\frac{1}{90}$. If the choice is m and n, then n and m is also acceptable. So the probability of Maryam picking the same counters as Hussain is $\frac{2}{90}$.

18.2 Frequency tree, given partial information

5   100 athletes are tested for taking unapproved drugs, and 15 test positive. 10 of the 100 athletes are in fact guilty of taking these drugs, and 90 are not. The test is always positive for athletes who are doping, but sometimes is falsely positive for an innocent athlete.

a   Complete the frequency tree in Figure 18.8.

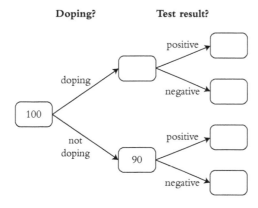

Figure 18.8

b   If I pick one of the innocent athletes at random, what is the probability they test positive?

Answer

5   a   First fill in that 10 athletes are doping. All of them will test positive, so we can fill in 10 and 0 in the top two final frequencies. Since a total of 15 test positive, 5 of the innocent athletes must test positive, allowing the tree to be completed (see Figure 18.9).

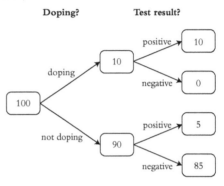

Figure 18.9

b   Out of 90 innocent athletes, 5 test positive. So the probability that a random innocent athlete tests positive is $\frac{5}{90} = \frac{1}{18}$.

18.3 Inverse problem: given a final probability of an event, solve for the initial conditions

This can be posed in a wide variety of ways.

6   Louis has 10 sweets in a bag, a mixture of toffees and chocolates. Sophia takes two sweets at random and eats them. The probability that Sophia ate two chocolates is $\frac{1}{3}$. How many chocolates were in the bag?

Answer

6   Suppose there are n chocolates. Then the probability that Sophia's first sweet is a chocolate is $\frac{n}{10}$. If it is a chocolate, the probability that Sophia's second sweet is a chocolate is $\frac{n-1}{9}$. So the overall probability of Sophia getting two chocolates is $\frac{n}{10} \times \frac{n-1}{9}$, which we are told is $\frac{1}{3}$. Therefore

$$\frac{n}{10} \times \frac{n-1}{9} = \frac{1}{3}$$
$$n(n-1) = 30$$
$$n^2 - n - 30 = 0$$
$$(n-6)(n+5) = 0$$

And so $n = 6$, since $n = -5$ is an absurd answer. There were 6 chocolates and 4 toffees.

7   A class of size n has a single set of twins. If I pick two students at random from the class, the chance that I pick the two twins is $\frac{1}{66}$.

   a   Show that n obeys the quadratic equation $n^2 - n - 132 = 0$.

   b   How many students are in the class?

Answer

   a   With n students in the class, the probability that the first student picked is a twin is $\frac{2}{n}$. If this happens, there are now $n - 1$ students left, of which one is the other twin. The probability that the second random choice is the matching twin is $\frac{1}{n-1}$. So the overall probability of picking the pair of twins is $\frac{2}{n} \times \frac{1}{n-1}$. Equating this to $\frac{1}{66}$ and rearranging gives the quadratic equation as follows.

$$\frac{2}{n} \times \frac{1}{n-1} = \frac{1}{66}$$
$$n(n-1) = 66 \times 2$$
$$n^2 - n - 132 = 0$$

   b   Since $n^2 - n - 132 = (n+11)(n-12) = 0$, there are either -11 or 12 students in the class. $n = 12$ is the only sensible answer.

8   In a box of raffle tickets, 10% of the tickets win a prize. If I buy two raffle tickets at random, the probability that I win two prizes is $\frac{1}{190}$. How many winning tickets are in the box?

Answer

8   Suppose there are n winning tickets in the box, out of a total of $10n$ tickets. The probability that the first ticket is a winner is $\frac{1}{10}$. If this happens, there are now $10n - 1$ tickets left, of which $n - 1$ are winners. So the probability that the second ticket is also a winner is $\frac{n-1}{10n-1}$. Therefore the overall probability of picking a pair of winning tickets is $\frac{1}{10} \times \frac{n-1}{10n-1}$. Equating this to $\frac{1}{190}$ and rearranging gives the equation $19(n-1) = 10n - 1$. Solving gives $9n = 18$, or $n = 2$. So there are only 2 winning tickets in the box, out of 20 altogether.

9   Katrin has a bag of sweets which are green or yellow in the proportion 1 : 2. She picks sweets at random to give to five of her friends: two get green sweets and three get yellow. The probability that the next friend gets a green sweet is now $\frac{2}{7}$. How many sweets did she start with?

9   Suppose she started with n green sweets and $2n$ yellow, so the total is $3n$. Then after she has given out 2 green and 3 yellow sweets, she has $n - 2$ green out of a total of $3n - 5$ sweets. The probability of the next friend getting a green sweet is therefore $\frac{n-2}{3n-5}$, which we are told is $\frac{2}{7}$, so that $\frac{n-2}{3n-5} = \frac{2}{7}$.

Therefore

$$7n - 14 = 6n - 10.$$

Solving the equation gives $n = 4$, and hence a total of 12 sweets in the bag, comprising 4 green and 8 yellow.

Inverse conditional probability – Bayes theorem

19.1 Enumeration of sets

1   50 students are asked about their families:

- 30 have at least one brother

- 25 have at least one sister

- 5 have neither a brother nor a sister.

   a   Represent this information on a Venn diagram.

   b   If I pick a student at random, what is the probability they have both a brother and a sister?

   c   If I pick a student at random and they have a brother, what is the probability they also have a sister?

Answer

1   a   The total number who have a brother **or** a sister is $50 - 5 = 45$. Since 30 students have a brother and 25 have a sister (giving a total of 55), there must be 10 who have a brother **and** a sister ($55 - 45 = 10$). This gives the Venn diagram in Figure 19.1.

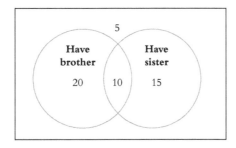

Figure 19.1

   Alternatively, let $x =$ the number with both a brother and a sister. So, since $30 - x$ have a brother but not a sister, and $25 - x$ have a sister but not a brother, we know that $(30 - x) + x + (25 - x) = 45$, and so $x = 10$.

   b   If I pick a student at random, the probability they have both a brother and a sister is $\frac{10}{50} = \frac{1}{5}$.

   c   30 students have a brother. If I pick one of them at random, the probability they also have a sister is $\frac{10}{30} = \frac{1}{3}$.

Part c illustrates the importance of being very clear what the reference class (from which we get the denominator) is. Here, it is the number of students who have a brother.

2   Consider Question 3 in Chapter 13, about the sports that students watch on TV. Suppose I pick a student at random and they watch rugby. What is the probability they also watch football?

Answer

2   The completed Venn diagram is reproduced in Figure 19.2.

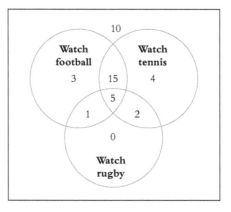

Figure 19.2

There are a total of $1 + 5 + 2 = 8$ students who watch rugby, and 6 of these also watch football. So if I pick a random student who watches rugby, the probability they also watch football is $\frac{6}{8} = \frac{3}{4}$.

19.2 Proportions and sets

3   70% of films shown in a chain of cinemas are from the USA. 60% of films shown are action films. Only 10% of films shown are neither from the USA nor action films.

a   Represent this information on a Venn diagram.

b   If I pick a film at random, what is the probability that this film is both from the USA and an action film?

c   If I pick an action film at random, what is the probability that it is **not** made in the USA?

Answer

3 a 90% of films are either made in the USA, action films, or both.
70% are made in the USA, 60% are action films, so 40% must be
both. This gives the Venn diagram shown in Figure 19.3.

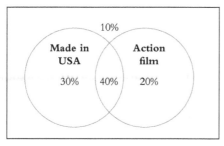

Figure 19.3

Alternatively, let x be the percentage which are both, and solve
the resulting equation: $70 - x + x + 60 - x = 90$, giving $x = 40$.

b 40% of films are both made in the USA and action films. So if I
pick a film at random, the probability that this film is both from
the USA and an action film is 0.4.

c 60% of films are action, and 20% are action and not made in the
USA. So if I pick an action film at random, the probability that it
is not made in the USA is $\frac{20}{60} = \frac{1}{3}$.

4 There are 40 fiction and 25 non-fiction books in a library. 80% of the
fiction books and 20% of the non-fiction books are paperback. If I
pick a paperback book at random, what is the probability it is fiction?

Answer

4 Create a frequency tree like that in Figure 19.4.

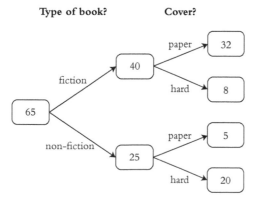

Figure 19.4

There are a total of $32 + 5 = 37$ paperback books. Of these, 32 are
fiction, so if I pick a paperback book at random, the probability that
it is fiction is $\frac{32}{37}$.

19.3 Expected frequency tree

5   A weather forecast is generally right. When it forecasts 'rain', 90% of
     the time it rains. When it forecasts 'no rain', 70% of the time it does
     not rain. In a typical September, 'rain' is forecast on $\frac{1}{3}$ of days and 'no
     rain' is forecast on $\frac{2}{3}$ of days.

   a   If I pick a random day in September, what is the probability that
        it will rain on that day?

   b   If it rains on a day in September, what is the probability that the
        forecast said it would rain?

Answer

5   a   Consider what we expect to happen during the 30 days of
          September. Using the information provided we can construct
          the expected frequency tree shown in Figure 19.5.

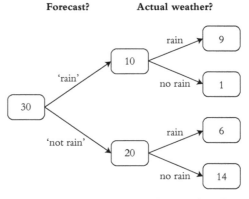

Figure 19.5

          We expect it to rain on 9 + 6 = 15 days, and so the probability
          that it will rain on a random day is $\frac{15}{30} = \frac{1}{2}$.

   b   Of the 15 days on which we expect it to rain, the forecast said
        it would rain on 9. So if it rains on a day in September, the
        probability that the forecast said it would rain is $\frac{9}{15} = \frac{3}{5}$.

6   A fair coin is flipped to decide whether your cricket team is going
     to bat first or second: Heads you bat first, Tails you bat second. If you
     bat first, your team wins 80% of the time. If you bat second, you win
     50% of the time.

   a   Out of 100 games, in how many do you expect to bat first?

b   Before you flip the coin, what is the probability of you winning the game?

c   If you win your match, what is the probability you batted first?

Answer

6   We can construct an expected frequency tree based on 100 games as shown in Figure 19.6.

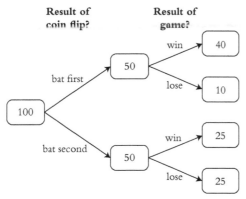

Figure 19.6

a   We expect to bat first in 50 games.

b   We expect to win in 40 + 25 = 65 games, so before we flip the coin, the probability of us winning the game is $\frac{65}{100} = 0.65$.

c   Of the 65 games we expect to win, we batted first in 40 of them. So if we win the match, the probability that we batted first is $\frac{40}{65} = \frac{8}{13}$.

Note that a common error is to say that out of 65 games that we expected to win, we batted first in *50* of them.

7   In an examination of 100 students, some are suspected of cheating in an exam but all deny it. They are wired up to a lie detector that will go 'ping' if it thinks the person is lying. The people who make the detector claim that if the person being tested is lying, there is a 90% chance the machine will go 'ping'. If the person being tested is genuinely not lying, there is a 10% chance the machine will get it wrong and go 'ping' anyway. Suppose 10% of the students cheat.

a   For how many students is the machine expected to go 'ping'?

b   If the machine goes 'ping', what is the probability that the student being tested has been cheating?

7 a We can construct an expected frequency tree, like that in Figure 19.7, based on 100 students.

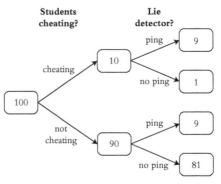

Figure 19.7

We expect the machine to go 'ping' for 18 students.

b Of the 18 students we expect to 'ping', 9 are truly cheating. So the probability that, when the machine goes 'ping', the student is truly cheating is $\frac{9}{18} = \frac{1}{2}$.

8 Opaque Bag A contains 4 blue and 1 green counters. Opaque Bag B contains 2 blue and 3 green counters. I pick a bag at random, take out a counter, check its colour and replace it.

a If I repeat this experiment 10 times, how many times do I expect to draw a blue counter?

b If I do the experiment once and draw a blue counter, what is the probability I have chosen Bag A?

Answer

8 The trick is to construct an expected frequency tree based on 10 repetitions of the experiment. This is shown in Figure 19.8.

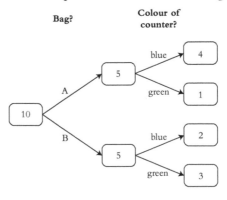

Figure 19.8

a Out of 10 repetitions, I expect to draw a blue counter 6 times.

b Out of 6 expected blue counters drawn, 4 are from Bag A. So if blue is drawn in a single experiment, the probability that it was from Bag A is $\frac{4}{6} = \frac{2}{3}$.

19.4 Probability trees

9 In my pocket I have a coin with both faces Heads, a coin with both faces Tails, and a fair coin. I pick a coin at random and, without looking at it, flip it. It lands with a Head up. What is the probability that I have flipped the coin with two Heads?

Answer

9 A probability tree is shown in Figure 19.9.

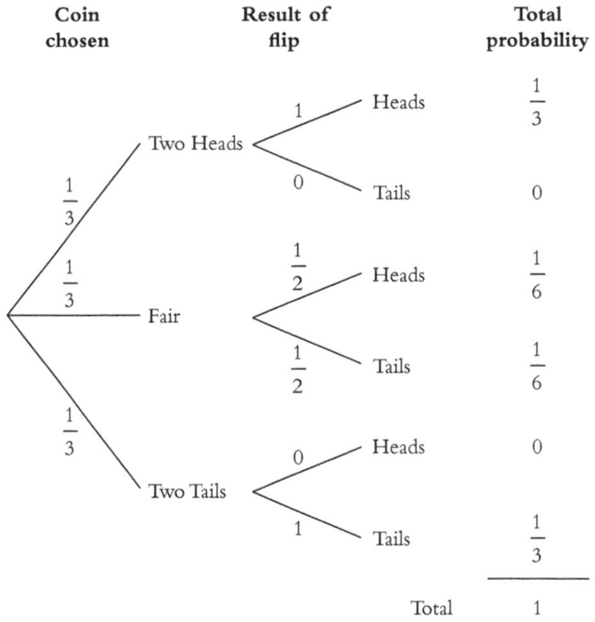

Coin chosen	Result of flip	Total probability

Figure 19.9

The overall probability of flipping a Head is $\frac{1}{3} + \frac{1}{6} = \frac{1}{2}$. The fraction of this total probability that is provided by 'two-Heads' is $\frac{1}{3} \Big/ \frac{1}{2} = \frac{2}{3}$. So the probability that the coin has two Heads is $\frac{2}{3}$.

However, it is much easier to build an expected frequency tree for 6 repeats of the experiment – this is shown in Figure 19.10.

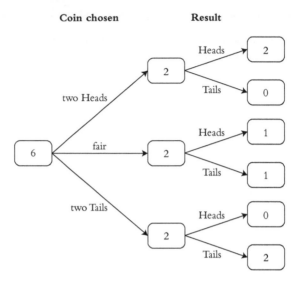

Figure 19.10

Of the three times the flipped coin comes up Heads, two are for the two-Headed coin. Therefore, after a single experiment that results in a Head, the probability that I have the two-Headed coin is $\frac{2}{3}$.

10   Opaque Bag A contains 1 blue and 2 green counters. Opaque Bag B contains 4 blue and 1 green counters. I pick a bag at random, and take out a counter. If I draw a blue counter, what is the probability I have chosen Bag A?

Answer

10   We can answer this using a probability tree like that shown in Figure 19.11.

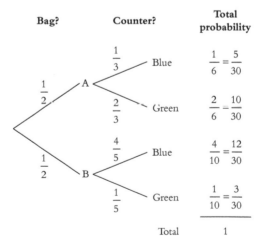

Figure 19.11

Note that the total probabilities have been expressed in terms of a lowest common denominator, 30. We see there is $\frac{5}{30} + \frac{12}{30} = \frac{17}{30}$ probability of picking a blue ball. Of this total probability, $\frac{5}{30}$ is provided by Bag A. Therefore the probability we have chosen Bag A is $\frac{5}{30} \Big/ \frac{17}{30} = \frac{5}{17}$.

The last piece of the argument is tricky, and could be expressed as follows: of 30 theoretical repeats of the experiment in which a blue ball was picked, 5 would have been from Bag A and 12 from Bag B. Therefore in a single experiment in which a blue ball was picked, the probability that we have picked from Bag A is $\frac{5}{17}$.

Alternatively, and as a check of the above, we can construct an expected frequency tree based on 30 repetitions of the experiment. This is shown in Figure 19.12.

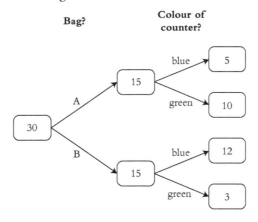

Figure 19.12

Out of 17 expected blue counters drawn, 5 are from Bag A. So if blue is drawn in a single experiment, the probability that it was from Bag A is $\frac{5}{17}$.

Although (in our opinion) the expected frequency produces a much more intuitive result, it is not always straightforward to select the appropriate number of experiments that gives all integer frequencies.

Introduction

This part contains a wide range of resources for further classroom activities about probability, including both classic ideas such as matching birthdays and the Monty Hall game, and more innovative activities on attitudes to risk based on 'behavioural economics' – the latter may appear to be more psychology than mathematics, but we believe the activities provide useful insights into uncertainty and its quantification.

There is, of course, far more material here than any one teacher could use in their classroom, given the limited time probability is allocated in the curriculum. However, we hope that teachers will be able to find resources suitable for supplementing the content that they need to teach, or indeed extension material that will intrigue their pupils. Some of the content might not be used in the classroom, but should still help a teacher to enhance their own knowledge of probability.

The chapters contain some material of a potentially sensitive nature to some individuals or religions, including discussions of gambling, eating pork, taking monetary risks, cancer, insurance, terrorism risk, congenital abnormalities, and family planning. Since this is a book for teachers rather than students, we have simply highlighted possible issues and leave the use of the material to teacher discretion.

References to further resources are given: we have tried to provide links to freely available online material and to avoid academic references; our exception is to include the journal *Teaching Statistics* as it contains so many relevant articles.

What's the best strategy?

20.1 Summary

There are a huge number of games that depend on chance and strategy, where players' approaches may depend on their attitude to risk and caution. Here are a few games that are not too complex, and do not involve playing cards or dice throws by students. They are intended to be reasonably fun to play, have a clear 'reveal' that indicates success or failure and which elicits a reaction, and should also illustrate particular aspects of probability and risk.

20.2 Possible classroom activities

THINK (SKUNK)

This is a popular game that involves some strategy and rather a lot of luck. We know it as *THINK* (it was originally termed *SKUNK*, but this repeats letters). Each student needs a scoring table (like Table 20.1), and there is one 'pair of dice' thrown for the whole class simultaneously – this would ideally be an electronic pair of dice on a whiteboard, but could be two spins by a chosen student on a dice spinner (Figure 20.1), or even a pair of real dice.

T	H	I	N	K

Table 20.1 Scoring table for THINK

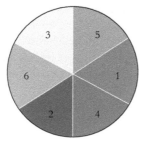

Figure 20.1 Dice spinner

Each letter of "THINK" represents a different round of the game, with each player recording points in each column of their own scoring sheet. In each round, the 'pair of dice' are repeatedly thrown and their combined score is accumulated in the appropriate column on the scoring sheet for players still in that round, with the following rules:

- If a '1' comes up on either 'dice', the player's go is over for that round, and all their points in that column are wiped out. Play then shifts to the next letter.
- If 'double 1' comes up, the player loses all the points accumulated in previous columns as well. Play then shifts to the next letter.
- If a '1' does not occur, players may choose either to try for more points on the next roll or to stop and 'bank' what they have accumulated. They then draw a line under their current accumulated score, which is 'safe', regardless of what else happens in that round (but may not be safe in subsequent rounds, if a 'double 1' occurs).

Scores are accumulated over the five rounds, corresponding to THINK.

Note: If a '1' or 'double 1' occur on the very first roll of a round, then that round is over and all players must take the consequences.

Some sources suggest that players stand to indicate they are still in a round, but they could just raise their hand when they stop playing.

Full teacher resources for *SKUNK* are available online [1]. Falk [2] discusses *THINK* in detail, for example pointing out that the probability of just one '1' is $\frac{10}{36}$, that of a 'double 1' is $\frac{1}{36}$, and that of a 'good roll' (both outcomes greater than 1) is $\frac{25}{36}$. The expected score in a good roll is 8.

THINK can result in a lot of groaning when scores are lost, but it illustrates the way that chance operates, and also allows people to choose between risk-taking and cautious strategies.

Play your cards right

Play Your Cards Right was a long-running TV game show in which large playing cards were revealed in sequence, and the contestants had to guess whether the next card would be higher or lower than the previous one. The exact probabilities can be worked out by enumeration, given the cards already exposed. This game has been suggested as a student activity [3, 4], but playing with full packs of cards does not seem straightforward. Hunt [5] suggests a simpler game using only six cards, or any objects numbered 1 to 6 that can be covered and then exposed, say counters with the numbers on one side only.

The game can be played in pairs (Players A and B), or in front of the class using large 'cards' numbered 1 to 6. The cards are arranged, face down, in two rows of three, the top row for Player A and the bottom row for Player B.

Player A turns over their first card, and then guesses whether the next card is higher or lower. They then guess whether their third card is higher or lower than their second. If both their guesses are right, they win the game immediately. Otherwise, Player B tries with their cards. In fact Player B only has to guess their second card, since A has revealed all their cards and so the number on Player B's final card can be worked out. Students should record how many times they win as Player A and as Player B and how many times nobody wins.

Hunt shows that if the players make sensible guesses (based on the most likely outcome), then Player A has $\frac{222}{360}$ chance of winning, and Player B $\frac{103}{360}$. Player A therefore has about a 2 to 1 advantage.

This game provides an example of calculating probabilities through simple enumeration of the remaining possibilities, and illustrates the importance of who goes first.

The price is right

The Price is Right is a TV game show that has been shown in many countries over a long period. There have been many class activities designed around this show, but here we consider one simple example.

Students play in pairs with the spinner in Figure 20.2 – the numbers are in the order of a standard dart-board.

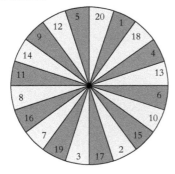

Figure 20.2 The Price is Right spinner

They take it in turns to be Player A or Player B, with the aim of getting the highest total score using either one or two spins. However, if the total goes over 20 the player scores zero. Player A spins first, and then Player B.

Since Player B knows Player A's score when they spin, this gives them some advantage and means their probability of winning should be greater than 50%. The critical decision is whether A should 'stick' after a spin of around 10, or spin again. Fletcher [6] discusses this in detail, working out the probability that A or B wins if they adopt different strategies, and concludes that Player A's strategy should be to spin again if they score 10 or less on their first spin – A then has a 46% chance of winning overall, while B's probability is 54%.

Shake and Take or Spin and Win

These games illustrate that increasing observations should lead to more confident estimates. They should be played in pairs, with students taking turns to be Player A and Player B. The two versions of the game illustrate the difference between sampling without replacement and with replacement.

Shake and Take: Player A puts five counters/cubes, each of which can be red or blue, into an opaque bag without Player B seeing, and then shakes the bag. Player B then has to guess how many red objects are in the bag, a number between zero and five. Player B takes out one object at a time, *without* replacement, and after each draw can, if they wish, stop the round and guess how many red objects were in the bag at the start – they only get one guess, but they can take up to five objects before they stop and guess. If Player B guesses correctly they score according to Table 20.2 – if they guess incorrectly they score nothing. There is clearly a judgement to be made about whether to stop early to try to get more points or, when there is less information, wait until confidence increases.

Spin and Win: Players should sit back to back. Player A chooses one of the six spinners in Figure 20.3, ensuring that Player B cannot see which one is chosen.

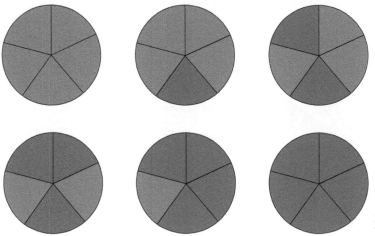

Figure 20.3 Spin and Win spinners

The aim is for Player B to guess which spinner is being used. Player A spins and announces the result for up to 5 spins. At any point Player B can guess which spinner is being used, but they only get one guess and the round is then over. Again, the scoring is shown in Table 20.2. This is essentially the same game as *Shake and Take*, but played using sampling with replacement.

Guess right after one draw/spin	5 points
Guess right after two draws/spins	4 points
Guess right after three draws/spins	3 points
Guess right after four draws/spins	2 points
Guess right after five draws/spins	1 points

Table 20.2 Scoring for Shake and Take and Spin and Win

The winner is the player with the highest score after, say, five rounds each.

The idea is to explore different strategies: is it better to be cautious and wait for more information, or make an early guess? The probability of the true number of red objects, or the true spinner, could in principle be calculated after each draw/spin *if* Player A is making their initial selection at random. In fact for *Shake and Take* there are five possible numbers of red objects after the first draw, four possible after the second draw and so on – if Player A has chosen the number of red objects at random, each of these possibilities is equally likely, and so the proposed scoring system means that the expected score is always 1 whatever Player B's strategy. However, in practice Player A will be using their own strategy in making their choice and may, for example, have a preference for all colours being the same. The number of objects, and the scoring system, can also be adjusted if it is felt it would make a better game.

20.3 Further questions and resources

There are many resources for games using chance, such as Math Academy's *Are You Game?* [7], *Probability Games from Diverse Cultures* [8], and *The Weakest Link* [9].

20.4 References

1   NCTM. *The Game of SKUNK* [Internet]. [cited 2015 Oct 29]. Available from: http://illuminations.nctm.org/Lesson.aspx?id=956

2   Falk R, Tadmor-Troyanski M. *THINK: A Game of Choice and Chance.* Teaching Statistics. 1999 Mar 1;21(1):24–7.

3   Fletcher M. *Play Your Cards Right.* Teaching Statistics. 1995 Jun 1;17(2):74.

4   DeRosa T. *High or Low: A Game of Probability* [Internet]. [cited 2015 Oct 29]. Available from: http://teaching.monster.com/training/articles/9164-high-or-low-a-game-of-probability

5   Hunt G. *Play Your Cards Right Again!* Teaching Statistics. 1996 Mar 1;18(1):15–6.

6   Fletcher M. *The Price is Right.* Teaching Statistics. 2005 Aug 1;27(3): 69–71.

7   Math Academy Probability Classroom Materials [Internet]. [cited 2015 Oct 29]. Available from: http://www.actuarialfoundation.org/programs/youth/materials-prob.shtml

8   McCoy L, Buckner S, Munley J. *Probability Games from Diverse Cultures* [Internet]. [cited 2015 Oct 29]. Available from: http://www.uccs.edu/Documents/pipes/mccoyprob-games.pdf

9   Fletcher M, Mooney C. *The Weakest Link.* Teaching Statistics. 2003 Jun 1;25(2):54–5.

What does 'random' look like?

21.1 Summary

Truly random sequences have no pattern and therefore cannot be condensed or summarised. If we wanted to describe the sequence, we would have to give them the entire list, since there is no simplifying rule for re-generating the entire sequence. But true randomness has some unexpected qualities – there are often more 'clusters' than we intuitively feel there should be. These exercises are intended to demonstrate the surprising properties of true randomness.

21.2 Possible classroom activities

Real or fake sequences?

Students should be on tables of at least four. Each student is given two strips like the one in Table 21.1. On the first strip, they write T or H underneath each number, pretending they are flipping a coin and writing T if a Tail or H if a Head. Then write 'fake' (lightly) on the back of the strip. For the second strip, flip a real coin 20 times, and write in each box T if a Tail or H if a Head. Write 'real' (lightly) on the back of the strip.

1	2	3	4	5	6	7	8	9	10	11	12	13	14	15	16	17	18	19	20

Table 21.1 Strips for real or fake sequences

Each table should now have at least 8 strips. Swap these with another table, face up so they cannot see which are real and which are fake sequences. Each table should then sort the strips into piles according to whether they think the sequences are real or fake. When instructed, turn them over and reveal the truth.

Students should, after some discussion, be able to work out fairly accurately which is real and which fake. What helps you choose which is real and which is fake? The crucial aspect is that the real sequences have longer runs or, equivalently, fewer switches between runs of Heads and runs of Tails. NRICH's *What Does Random Look Like?* [1] features a simulation for flipping coins and looking at run lengths – different-length sequences can be chosen. Simulating 10 000 sequences of length 20, and recording the longest run in each, gave the bar chart in Figure 21.1. In nearly half the sequences there is a run of at least five, and only around 20% do not have a run of at least four.

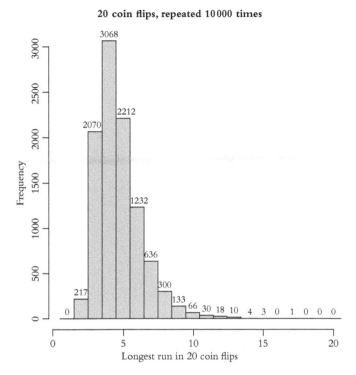

20 coin flips, repeated 10 000 times

Figure 21.1 Bar chart of runs in coin flips

Ask everyone in the class to flip a coin 11 times. It is more likely than not that each student will get a run of at least four Heads or Tails, so around half the class should get a run of at least four. Why should it not be surprising to get a run of four in 11 flips of a fair coin?

In a sequence of 11 flips, there are eight runs of four flips (1-2-3-4, 2-3-4-5, ... , 8-9-10-11). In any *particular* run of four flips, the probability of getting four Heads is $\frac{1}{2^4} = \frac{1}{16}$. The probability of getting four Tails is also $\frac{1}{16}$, and so the probability of getting a run of four is $\frac{2}{16} = \frac{1}{8}$. So there are eight runs, each with probability $\frac{1}{8}$ of being a run of four, so the expected number of runs of four is 1. (Note that the sequences of four are certainly not independent as they overlap, but the expected total still holds. This is due to the general result, which is taught in more advanced courses, that the expected sum of a set of random quantities is the sum of their individual expectations, even if they are not independent.) Since the expected number of runs of four is 1, we should not be surprised to get a run of four – although we worked out the probability (51%) using computer simulation rather than clever maths, which might seem a bit like cheating.

Random or patterns?

In three of the four pictures of 9 × 9 grids of colours in Figure 21.2 the colours were chosen at random from a palette of nine colours. In one of the pictures, the colours have been carefully chosen to form a very special pattern. Which is the non-random one? What pattern does it have?

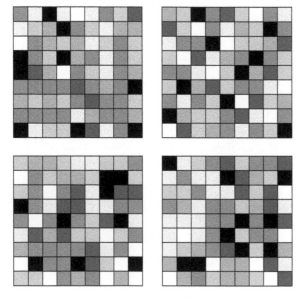

Figure 21.2 Random or not?

The non-random one is in the top-right. Why is this clear? The others all have some clustering, with same colours next to each other. The top-right picture is too regular to be random. Careful observation should reveal that the non-random one is a solution to a Sudoku puzzle, and a very special one too. See Understanding Uncertainty's *Pure Randomness in Art* for a full explanation [2].

Does your shuffle look random?

The shuffle facility on music players provides a fine example of where we experience the oddities of randomness in real life. Unfortunately, direct exploration of students' playlists is not appropriate as a classroom activity. A class could, however, be asked about their impressions of the shuffle facility, and may reflect that tracks from a particular album appear to repeat quickly, while some albums never seem to play; alternatively, students could be asked, as homework, to go through 20 tracks on a shuffle (they can fast forward these) and see how many repeated artists and albums there are. This can be correlated with the number of tracks on their music list (and of course whether they only feature one favourite artist!).

It is possible to conduct a simulation in class, following the approach of Slauson [3].

Prepare a total track list of 100 albums of 10 songs each [*or use the one on our website*]. Each person then creates a random playlist of 20 tracks (around an hour) by generating 20 random numbers between 1 and 1000 (say using Excel or a 0–9 spinner for each digit), and finding the relevant songs from the track list. How many of the playlists have repeated albums? It would be possible (but less memorable!) to observe the same results simply by generating 20 numbers between 1 and 1000 and observing whether there are pairs of numbers with less than 10 between them, or are in the same 'decade', say 310 to 319.

It is rumoured that Apple changed the iPod shuffle so that it was impossible for the same track to come up twice in a row – they made it less random so that it appeared more random. The exact probability of a matching album follows from the formula and spreadsheet given in *Happy birthday to you, and you, and …* (Chapter 31). The event that each of 20 random tracks is from a different one of 100 albums is essentially the same as 20 people choosing different numbers between 1 and 100 (the number of tracks on each album is irrelevant, as long as it is the same for each album). So if 20 songs are chosen at random from a music list of 100 albums, there is a rather remarkable 87% chance that an album will be repeated.

21.3 Further reading and resources

NRICH features a 'coin toss checker' that will generate random flips, allow you to change some, and then assess whether the sequence is truly random or not [4].

Children's perception of randomness is explored by Green [5] using patterns of counters on grids and asking whether the children thought they looked random.

The randomness of the iPod shuffle has been discussed in newspaper articles [6], and by Ziegler and Garfield [7, 8] as a basis for getting students to think of tests of randomness.

21.4 References

1   NRICH. *What Does Random Look Like?* [Internet]. [cited 2015 Oct 16]. Available from: http://nrich.maths.org/7250

2   Understanding Uncertainty. *Pure Randomness in Art* [Internet]. [cited 2015 Oct 16]. Available from: http://understandinguncertainty.org/node/1066

3   Slauson L. *Is the iPod shuffle feature truly random? A simulation activity.* [Internet]. [cited 2015 Oct 16]. Available from: https://www.causeweb. org/webinar/activity/2009-06/

4   NRICH. *Coin Toss Checker* [Internet]. [cited 2015 Oct 23]. Available from: https://nrich.maths.org/6078

5   Green D. *Recognising randomness.* Teaching Statistics. 1997 Jun 1;19(2):36–9.

6   Bialik C. *How Random Is the iPod's Shuffle?* [Internet]. [cited 2015 Oct 16]. Available from: http://www.wsj.com/articles/ SB115876952162469003

7   Ziegler L, Garfield J. *Exploring students' intuitive ideas of randomness using an iPod shuffle activity.* Teaching Statistics. 2013 Mar 1;35(1):2–7.

8   Ziegler L. *How Random is the iPod's Shuffle?* [Internet]. [cited 2015 Oct 16]. Available from: https://www.amstat.org/education/stew/ pdfs/HowRandomIsTheiPodsShuffle.pdf

How should we change our beliefs?

22.1 Summary

This section explores how our beliefs should be changed by additional evidence. This concerns conditional probability, using the mathematical result known as **Bayes theorem**, after the Reverend Thomas Bayes, a non-conformist minister who died in 1761.

Bayes theorem can be expressed as follows. Let A and B be two events. Then

$$P(B \mid A) = \frac{P(A \mid B)\ P(B)}{P(A)}$$

where $P(B \mid A)$ represents the 'probability of B, given A has occurred'. In words, the formula shows how our initial belief in B, as measured by $P(B)$, is revised to a new belief $P(B \mid A)$ on receipt of the evidence A. So Bayes theorem is a mathematical representation of learning from new information. This is also why this area is sometimes known as 'inverse' probability – we start with the conditional probability $P(A \mid B)$, and then we end up with $P(B \mid A)$, the 'inverse'. We show below how this is represented by a re-ordering on a tree.

Bayes theorem is trivial to prove. We know $P(A \text{ and } B) = P(A \mid B)P(B)$ – the rule of joint probability for dependent events. We use this every time we multiply along branches of a probability tree, since if B occurs first, $P(B)$ is the first probability, and if A occurs second, then it is conditional on B having already occurred, so the second probability is $P(A \mid B)$. But $P(A \text{ and } B) = P(B \text{ and } A)$ as the order is irrelevant, and $P(B \text{ and } A) = P(B \mid A)P(A)$, and so it follows that $P(B \mid A)P(A) = P(A \mid B)P(B)$.

Dividing each side by $P(A)$ gives the theorem.

We have already seen in Chapter 1 how a complex inverse probability problem involving doping in sports can be solved in a fairly straightforward way using expected frequency trees, with numerous worked examples in Chapter 19. Here we explore some of the further issues involved. Students should have the skills to handle these difficult and yet extremely practical and relevant questions, if they have worked through the analysis at the higher level in *The dog ate my homework!* in Chapter 6, or some of the examples in Chapter 19.

A running theme is the need to be obsessively pedantic about conditional probability statements, and how an expression of the form 'out of 100 Xs, we would expect Y to occur in 20 of them' can bring clarity to potentially confusing statements about the probability of Y.

22.2 Possible classroom activities

Doping in sports

Suppose a screening test for doping in sports is claimed to be '95% accurate', meaning that 95% of dopers, and 95% of non-dopers, will be correctly classified. Assume 1 in 50 athletes are truly doping at any time. If an athlete tests positive, what is the probability that they are truly doping?

Before doing the calculations, discuss what students think the answer might be.

This example featured in Chapter 1, where we pointed out that the best approach is through an expected frequency tree.

There is a problem in deciding how big to make the population under consideration. The best values are those that allow you to work in integers throughout; 1000 is always a good place to start. Figure 22.1 starts with 1000 athletes, of whom 20 are doping. All but one of them are detected, but 49 non-dopers also have positive tests.

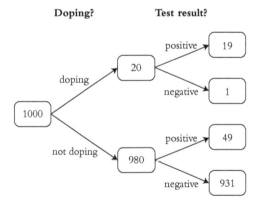

Figure 22.1 Expected frequency tree showing detection of athletes who are doping

We expect a total of 19 + 49 = 68 positive tests, of whom only 19 are truly doping. So if someone tests positive, there is only a $\frac{19}{68}$ = 28% chance they are truly doping. The majority (72%) of positive tests will result in false accusations.

One way of thinking of this process is that we are 'reversing the order' of the tree, as shown in Figure 22.2.

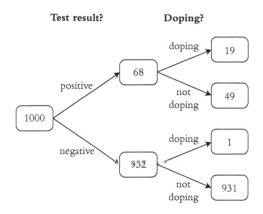

Figure 22.2 Reversed expected frequency tree

This reversed tree arrives at exactly the same numbers for the final outcomes, but in the order in which we find the information (testing and then finding the truth about doping), rather than the timeline on which events actually occurred (doping and then testing).

Note that we have to be *extremely* careful about the wording of the question, in defining what we mean by '95% accurate'. We do *not* mean that 95% of the test results are correct – we have just shown this is not true. We must be clear that we are providing the probability of the correct test result, *given* that someone is doping, or *given* that they are not doping. Many such questions use imprecise and confusing wording.

John Haigh discusses sports doping in *Plus* magazine [1], which includes a useful animation from the Understanding Uncertainty website [2].

Cancer screening

> We emphasise that any discussion of cancer, and perhaps particularly breast cancer, must be handled with care due to possible family experiences.

A standard example for this type of analysis is in screening for diseases such as breast cancer. The analysis precisely follows that of doping in sports.

> Women over 50 in the UK are routinely offered screening for breast cancer using an X-ray called a mammogram. Those with a positive mammogram are recalled for further investigations. It has been estimated that mammography will detect approximately 90% of breast cancers. It has also been estimated that around 10% of women with no cancer will still receive a positive result. Suppose that around 10 out of 1000 women being screened have breast cancer.
>
> If a woman has a positive mammogram, what is the probability she truly has breast cancer?
>
> Before doing the calculations, discuss what the answer might be.

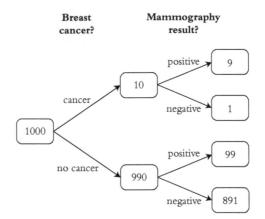

Figure 22.3 Expected frequency tree for breast cancer screening

Breast cancer? — Mammography result?

1000 → cancer 10 → positive 9, negative 1

1000 → no cancer 990 → positive 99, negative 891

Figure 22.3 shows that of 9 + 99 = 108 positive tests, only 9 are from women who truly have cancer, so if a woman has a positive mammogram, the chance she has cancer is around $\frac{9}{108}$ = 8%.

In fact the information leaflets for breast screening in the UK report that around 1 in 4 positive mammograms actually turn out to be cancer, which is higher than 8% but still in the minority.

This example is covered on the Understanding Uncertainty website, which features a helpful animation [3].

The taxi-cab problem

A classic psychology experiment goes as follows.

'A taxi-cab was involved in a hit and run accident last night, and a witness identifies the taxi-cab as blue. The court tests the reliability of the witness under the circumstances that existed on the night of the accident and concluded that the witness could correctly identify each of the two colours 80% of the time but would fail 20% of the time. In the city, 85% of the cabs are green and 15% are blue. What is the probability that the cab involved in the accident was blue rather than green?' [4].

Before doing the calculations, discuss what the answer might be.

By now students should know how to handle such a problem, by imagining what we would expect to happen in 1000 such situations (see Figure 22.4).

The witness would identify 120 + 170 = 290 taxi-cabs as 'blue', of which 120 were truly blue, so the probability in this particular case, that the cab is truly blue, is $\frac{120}{290}$ = 41%. (Note that since all the final numbers divide by 10, we could have used 100 as the initial number of cabs.)

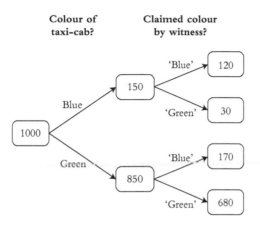

In numerous experiments people respond with a high probability such as 80%, apparently ignoring the information about the comparative rarity of blue cabs. This is known as the 'base-rate' fallacy.

Bedwell [5] explores this problem as a class activity, suggesting questions such as 'How do you imagine the tests were carried out to determine the reliability of the witness?'

22.3 Further reading and resources

See Wikipedia [6] and *Plus* magazine [7] for more problems solved with Bayes theorem. However, they both use probability rather than expected frequencies, which makes it all rather more difficult.

22.4 References

1   Haigh J. *The logic of drug testing* [Internet]. [cited 2015 Oct 28]. Available from: https://plus.maths.org/content/logic-drug-testing

2   Understanding Uncertainty. Bayes theorem simulation [Internet]. [cited 2015 Oct 28]. Available from: http://understandinguncertainty. org/files/animations/BayesTheorem1/BayesTheorem.html

3   Understanding Uncertainty. *Screening for breast cancer* [Internet]. [cited 2015 Oct 28]. Available from: http://understandinguncertainty.org/node/182

4   Kahneman D. *Thinking, Fast and Slow.* Farrar, Straus and Giroux; 2011. p. 511.

5   Bedwell M. *Slow thinking and deep learning: Tversky and Kahneman's taxi cabs.* Teaching Statistics. 2015 Sep 1;37(3):85–8.

6   Wikipedia. *Bayes' theorem* [Internet]. [cited 2015 Oct 28]. Available from: https://en.wikipedia.org/wiki/Bayes%27_theorem

7   K.E.M *Ye banks and Bayes* [Internet]. Plus magazine [cited 2015 Oct 28]. Available from: https://plus.maths.org/content/ye-banks-and-bayes

Chapter
23

How probable is probable?

23.1 Summary

We all use a wide variety of terms to indicate uncertainty – 'could', 'maybe', 'possible' and so on. Can these be interpreted as numerical probabilities? These activities are intended to show how difficult this is, as the interpretation of these terms varies widely between individuals and contexts. This has not stopped many organisations trying to standardise verbal expressions of uncertainty, and some of these efforts are described here for comment and criticism.

23.2 Possible classroom activities

Interpretation of words

Read the following paragraph [*available on website*].

> Arthur was worried. It was almost certain there would be a maths test today, and he hadn't been paying much attention recently. Sally would probably get more marks than him, but there was a distinct possibility that Zak would mess up. The weather forecast said it might rain, so he took a coat, and as he walked to school he thought he was likely to meet Zak, who always played around, and could make him late. If he were late, he was certain to get into trouble. Perhaps there would be a fire drill to disrupt the test? But really there was little chance of that, and it was also extremely unlikely an asteroid would hit the school. It was going to be a difficult sort of day.

Underline the words or phrases that express uncertainty, such as 'could', 'likely' and so on. Make a list of these words, and rank them in terms of highest to lowest probability. Put each word on the vertical probability scale in Figure 23.1, for example, if you think that 'almost certain' is near 50%, write it next to 50%.

Collect together the responses and discuss the ranges in opinion.

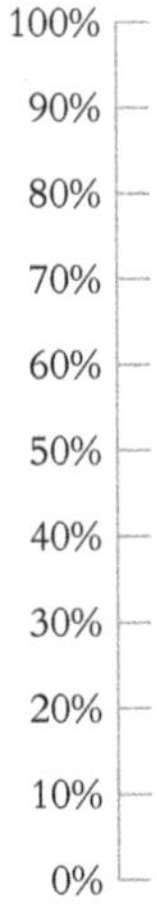

100% —
90% —
80% —
70% —
60% —
50% —
40% —
30% —
20% —
10% —
0% —

Figure 23.1 Vertical probability scale

There should be reasonably good agreement for terms like 'extremely unlikely', but 'could' could mean almost anything.

Drug side effects

Most medicines have occasional side effects – the drugs may make some people drowsy, make their muscles ache and so on. If you were told a side effect, say headache, was 'common', how frequently do you think it would occur, as a percentage of people taking the medicine? What if the headaches were described as 'very common'?

The official European Medicines Agency scale defines 'common' as a frequency between 1% and 10%, and 'very common' as anything above 10% [1].

What do you think of this?

This seems an example where the 'official' definition of a word has little resemblance to its use in everyday language, and could lead to patients believing side effects occur far more often than intended.

Amniocentesis

This discusses congenital abnormalities in children. Teacher discretion is advised.

Pregnant women usually have a screening test for possible problems with their foetus. A test result that shows any probability above 1 in 150 (0.6%) of having a baby with Down's syndrome is called a 'higher-risk result' on the NHS Choices website [2]. Such women are offered an amniocentesis to confirm or rule out the diagnosis, but this procedure carries some risk of causing a miscarriage – this risk is estimated to be about 1%, and is described as a 'small associated risk' by NHS Choices [3]. What do you think of this? Why do you think this wording has been used?

Note that here a 'higher risk' is numerically smaller than a 'small risk'. The wording may have been chosen to encourage women to have an amniocentesis if they have a positive screening test, and not worry too much about the risk of miscarriage.

IPCC scale

Many organisations have tried to standardise terms expressing uncertainty.

Table 23.1 shows the scale used by the Intergovernmental Panel on Climate Change (IPCC).

Verbal expression	Numerical probability range
virtually certain	99–100%
extremely likely	95–100%
very likely	90–100%
likely	66–100%
more likely than not	50–100%
about as likely as not	33–66%
unlikely	0–33%
very unlikely	0–10%
extremely unlikely	0–5%
exceptionally unlikely	0–1%

Table 23.1 IPCC scale

For example, in their recent report, they claimed that 'It is extremely likely that human influence has been the dominant cause of the observed warming since the mid-20th century.' [4] What do you think of this scale? How does it fit with the numbers assessed by the class in the activity at the start of this chapter?

Terrorism alerts

This discusses possible terrorist attacks. Teacher discretion is advised.

The UK has a scale to assess the current level of threat from international terrorism. The MI5 website states [5]:

Threat levels are designed to give a broad indication of the likelihood of a terrorist attack.

- *LOW means an attack is unlikely.*
- *MODERATE means an attack is possible, but not likely.*
- *SUBSTANTIAL means an attack is a strong possibility.*
- *SEVERE means an attack is highly likely.*
- *CRITICAL means an attack is expected imminently.*

What do you think these phrases mean in terms of probability?

When the level was raised to *SEVERE,* the Home Secretary said 'The Joint Terrorism Analysis Centre has today raised the threat to the UK from international terrorism from *SUBSTANTIAL* to *SEVERE.* This means that a terrorist attack is highly likely, but I should stress that there is no intelligence to suggest than an attack is imminent.'

What does 'highly likely' mean here?

One problem is that no time horizon is provided to assess what 'highly likely' means. Next week? Next month? The next 10 years?

23.3 Further reading and resources

Wikipedia discusses the use of verbal terms by the intelligence community. [6] The CIA's involvement is also covered in a newspaper article, which illustrates the range of opinions held by intelligence experts [7].

23.4 References

1   European Medicines Agency. *Section 4.8: Undesirable effects* [Internet]. [cited 2015 Oct 20]. Available from: http://www.ema.europa.eu/docs/en_GB/document_library/Presentation/2013/01/WC500137021.pdf

2   NHS Choices. *Screening for Down's, Edwards' and Patau's syndromes* [Internet]. [cited 2015 Oct 20]. Available from: http://www.nhs.uk/conditions/pregnancy-and-baby/pages/screening-amniocentesis-downs-syndrome.aspx

3   NHS Choices. *Amniocentesis* [Internet]. [cited 2015 Oct 20]. Available from: http://www.nhs.uk/conditions/Amniocentesis/Pages/Introduction.aspx

4   Intergovernmental Panel on Climate Change (IPCC). *Climate Change 2013: The Physical Science Basis. Summary for Policymakers* [Internet]. 2013 [cited 2015 Oct 20]. Available from: https://www.ipcc.ch/pdf/assessment-report/ar5/wg1/WG1AR5_SPM_FINAL.pdf

5   MI5 – The Security Service. *Terrorist threat levels* [Internet]. [cited 2015 Oct 20]. Available from: https://www.mi5.gov.uk/threat-levels

6   Wikipedia. *Words of estimative probability* [Internet]. [cited 2015 Oct 20]. Available from: https://en.wikipedia.org/wiki/Words_of_estimative_probability

7   Arnett G. *How probable is 'probable'?* [Internet]. The Guardian [cited 2015 Oct 20]. Available from: http://www.theguardian.com/news/datablog/2015/aug/14/how-probable-is-probable

Misconceptions

24.1 Summary

There are some common mistakes made in working with probability. In this chapter we consider a set of statements that invite class critique and discussion. Many of these involve a misinterpretation of either the *event*, or the *circumstances* of the event. We should always think carefully about:

- What is the precise event of interest?
- What are the conditions and the subjects under consideration?

24.2 Possible classroom activities

Snakes on a plane – conditional probability

> When you are flying, always take a pet snake with you in your hand luggage. The probability of there being *two* snakes on the plane is almost zero, so you will be safe from snake attack. Sensible claim?

There are two levels of idiocy in this statement. The first is that, by taking a snake on a plane, it is *guaranteed* there is at least one snake on the plane. Second, and more important, this is a confusion of *conditional* and *unconditional* probability. It is true that, overall, there is a very small *unconditional* probability of two people choosing to take snakes on a plane. However, the relevant probability is the *conditional* probability of someone else taking a snake, given you have already done so. Since we assume snake-carrying people are independent, this conditional probability is unaffected by your own snake.

This so-called paradox is often stated using a bomb instead of a snake.

How many Heads?

> When flipping 10 coins, you are just as likely to get 1 Head as 5 Heads. Sensible claim?

Each actual sequence of 10 flips is equally likely, but there are far more sequences with 5 Heads ($^{10}C_5 = 252$) than there are with 1 Head ($^{10}C_1 = 10$). This is essentially a misinterpretation of the event of interest – we want the probability of a summary of a sequence, rather than a particular sequence. (See Chapter 25: *Heads or Tails, boy or girl?*)

Boys and girls

> The probability of having a boy baby is $\frac{1}{2}$. In a family of six children, the sequence BGGBGB is more likely than BBBBGB. Sensible claim?

The first sequence seems more 'representative', and the run of four boys in the second sequence appears rare, but actually these two sequences are equally likely. The crucial distinction is between the chances of these two *specific* sequences, which are equal, and the chance of getting 3 boys rather than 5 boys, which is certainly higher. This is essentially a misinterpretation of the event of interest.

As a discussion point, the probability of a baby being a boy is actually around 51%, slightly more than $\frac{1}{2}$, for reasons that are unclear. (See Chapter 25: *Heads or Tails, boy or girl?*)

Adding or multiplying?

> The chance of getting a six when throwing a die is $\frac{1}{6}$. So if I throw it twice, the chance of getting at least one six is $\frac{2}{6} = \frac{1}{3}$. Sensible claim?

The correct probability of getting at least one six is
$1 - $ probability of getting no sixes $= 1 - \left(\frac{5}{6} \times \frac{5}{6}\right) = \frac{11}{36}$,
which is slightly less than $\frac{1}{3}$. In the question box, the probabilities have been added rather than their complements multiplied and then subtracted from 1. One way to show this cannot be correct is to extend the reasoning to 12 throws of the die – the probability of getting at least one six cannot be $12 \times \frac{1}{6} = 2$.

However, 'addition' *is* approximately right if the chances are small, and this can cause confusion. Suppose, for example, we are talking about correctly guessing a random number between 1 and 100. The chance of a correct guess at the first attempt is $\frac{1}{100} = 0.01$. If two guesses are made, the exact chance of getting at least one right is $1 - \left(\frac{99}{100} \times \frac{99}{100}\right) = 0.0199$, which is very close indeed to the 'addition' result of $\frac{2}{100} = 0.02$. So for small probabilities of rare events, adding to get the overall probability of at least one of them happening will give an approximately correct answer.

Gambler's fallacy

> If you flip a fair coin and it comes up 6 Heads in a row, it is more likely to come up Tails next time. Sensible claim?

The crucial word is 'fair' – this means flips are independent and the probability of either a Head or a Tail is always exactly $\frac{1}{2}$. There is no memory in the system, and no possibility of learning from experimental data. Believing that a particular event is due is known as the 'gambler's fallacy'. In fact, if there is any suspicion that the coin is not fair, it would be more reasonable to say it is more likely to continue the run and be Heads next time as well, as the coin may be biased in favour of Heads.

Chance of a draw

In a football match, a team will either win, lose or draw, so the probability of each is $\frac{1}{3}$. Sensible claim?

Just because there are three possible results does not mean they are equally likely. In fact in top-class football, the home team wins in about 45% of matches, the away team wins in about 30%, and about 25% are a draw.

Completing the pile

Two piles of bricks, A and B, have 4 and 3 bricks respectively. If bricks are added to the piles at random, the two piles are equally likely to be first to reach 5 bricks. Sensible claim?

The reasoning behind this (incorrect) answer might be: 'The probability of a brick in each pile is $\frac{1}{2}$. As A only needs one more brick, then it has probability $\frac{1}{2}$ of being finished first.' Another 'intuitive' (but incorrect) response is to say that there are three possibilities that can lead to a finished pile: A, BA and BB. Pile A is completed first in two out of three cases and therefore the probability of A being finished first is $\frac{2}{3}$. However, this assumes these three events are equally likely. To investigate further, it is a good idea to draw a probability tree (Figure 24.1).

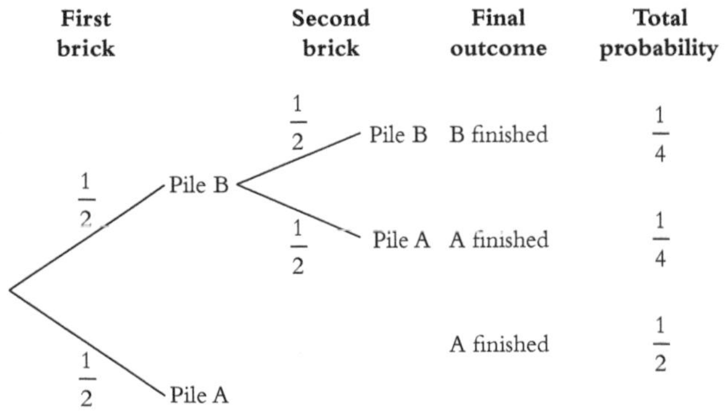

Figure 24.1 Probability tree showing completion of the two piles of bricks

First brick	Second brick	Final outcome	Total probability
	$\frac{1}{2}$ Pile B	B finished	$\frac{1}{4}$
$\frac{1}{2}$ Pile B	$\frac{1}{2}$ Pile A	A finished	$\frac{1}{4}$
$\frac{1}{2}$ Pile A		A finished	$\frac{1}{2}$

This reveals that the probability that A is finished first is $\frac{3}{4}$.

This question is based on the problem of how to allocate points in a game that has to terminate before it is complete. Discussion of this problem by Pascal and Fermat was the origin of the mathematics of probability.

Learning from experience

One hundred drawing pins are dropped. 68 land 'up' and 32 land 'down'. When the experiment is repeated, these results are all equally likely:

- 36 'up' and 64 'down'
- 63 'up' and 31 'down'
- 51 'up' and 49 'down'. Sensible claim?

This could be an example of the misconception that if three possible results are listed, they are all equally likely. Or the third option might be thought more likely, by incorrectly reasoning that there are two possible outcomes for the drawing pins, and hence the chances of either must be 50–50 (see *Chance of a draw* above). The first option might be considered more likely, by incorrectly thinking that future observations should somehow 'balance out' the past (see *Gambler's fallacy* above).

After reflection, the second option should appear more likely, as it correctly learns from the experience from the first set of data.

Independence

The famous mathematician Hardy took an umbrella to cricket matches, since if you forget your umbrella it is more likely to rain. Sensible claim?

It only appears more likely to rain when we forget our umbrellas because we notice such unfortunate events. The events 'rain later' and 'carry umbrella' are independent and so the chance of one is not influenced by the chance of the other.

Some people, however, have a strong aversion to feeling regret, and so for them it is perfectly sensible to be cautious and carry an umbrella at all times.

24.3 Further reading and resources

Many of these examples are adapted from NRICH's *Do You Feel Lucky* [1], Monks [2], and the fine collection of slides provided by Dan Walker [3]. See also *Lottery myths* in Chapter 27: *It's a lottery* and the Monty Hall problem in Chapter 28: *Switch or stick?*

24.4 References

1   NRICH. *Do You Feel Lucky?* [Internet]. [cited 2015 Oct 24]. Available from: http://nrich.maths.org/7222

2   Monks AR. *Equally Likely*. Teaching Statistics. 1985 Sep 1;7(3):66–9.

3   Walker D. *Probability* [Internet]. TES Resources. [cited 2015 Oct 24]. Available from: https://www.tes.com/teaching-resource/probability-6321017

Chapter 25

Heads or Tails, boy or girl?

25.1 Summary

Permutations and combinations are not really part of probability at all – they simply provide a way of enumerating possibilities. But the combination formula can be interesting in its own right, as well as being useful for calculating probabilities. Using a simple model for coin flipping, we develop Pascal's triangle and the formula for combinations.

25.2 Possible classroom activities

See the accompanying sheets on the website for supporting material.

Heads or Tails

Experimental distribution

Pairs of students should flip a coin, or spin a 50–50 spinner marked Heads and Tails, 10 times and count how many Heads result. Repeat this experiment 10 times. Then construct a bar chart of the results, showing how often each of 0, 1, 2, …, 9, 10 Heads came up. Estimate the probability of getting exactly 5 Heads and 5 Tails just using this data. Finally, pool data across the class to produce a 'smoother' experimental distribution, calculate the proportion for each number of Heads, and use this to estimate the probability of getting exactly 5 Heads and 5 Tails.

Note that the chance of getting 10 Heads or 10 Tails is $\frac{1}{2^{10}} = \frac{1}{1024}$, or very close to 1 in a thousand; this is a useful fact to remember.

Look at the Lightning animation on the website to see what happens with 20 coin flips.

The chance of getting 20 Heads or 20 Tails is $\frac{1}{2^{20}} = \frac{1}{1\,048\,576}$ or very close to 1 in a million.

If this experiment were repeated many thousands of times, what would the shape of the bar chart tend to? This is quite tricky, and we need to use some theory.

Theoretical distribution

We want the probability of getting, say, exactly 5 Heads in 10 flips. Each specific sequence of Heads and Tails is equally likely, but some have more Heads than others, so we need to count how many of these unique sequences have 5 Heads, and divide it by the total number of possible sequences to find the probability.

$$\text{Probability of exactly 5 Heads} = \frac{\text{Number of sequences with 5 Heads}}{\text{Total number of possible sequences}}.$$

We therefore would like a general formula, for any number of flips, for both the total number of possible sequences and the number containing 0, 1, 2, … Heads. Each pair of students should explore this for 1, 2, 3 and 4 flips, completing Table 25.1.

Number of flips	Unique sequences of Heads and Tails
1	H
	T
2	HH
	HT
	TH
	TT
3	HHH
	HHT
	HTH
	HTT
	THH
	THT
	TTH
	TTT
4	…

Table 25.1 Record of coin flips

They should then complete Table 25.2.

	Number of different sequences with this number of Heads						
Number of flips	0	1	2	3	4	5	…
1	1	1	X	X	X	X	X
2	1	2	1	X	X	X	X
3	1	3	3	1	X	X	X
4	…	…	…	…	…	X	X

Table 25.2 For 1, 2, 3, 4, … flips, the number of different sequences with 0, 1, 2, … Heads

Spot the pattern! What do you predict would happen for 5 flips, or 6 flips?

The first column is always 1, then each entry is the sum of the entry directly above and above to the left. Adding up the counts for each row, we see that they total 2^n, where n is the number of flips. So, what is the theoretical probability of flipping 2 Heads in 3 flips? ($\frac{3}{8}$) What about 5 Heads in 10 flips? (If the table is completed to 10 flips, you should find that this theoretical probability is $\frac{252}{1024} \approx 25\%$. So there is almost exactly a 1 in 4 chance of getting exactly 5 Heads in 10 flips.)

The numbers in Table 25.2 give Pascal's triangle, which has many wonderful properties. The first six rows are:

$$
\begin{array}{ccccccccccc}
 & & & & & 1 & & & & & \\
 & & & & 1 & & 1 & & & & \\
 & & & 1 & & 2 & & 1 & & & \\
 & & 1 & & 3 & & 3 & & 1 & & \\
 & 1 & & 4 & & 6 & & 4 & & 1 & \\
1 & & 5 & & 10 & & 10 & & 5 & & 1
\end{array}
$$

The *Lightning* animation will tend to the shape of the row of Pascal's triangle corresponding to 20 flips.

The formula for Pascal's triangle

The numbers in Pascal's triangle may be obtained by a sequential process, and from this we can derive the formula for combinations – we shall use this approach again when calculating the chances of winning a lottery (see Chapter 27).

Suppose we want the number of different sequences of 10 flips that contain 5 Heads. Imagine the 10 coin flips in a row, labelled 1 to 10, 5 of which are Heads, but we don't know which. Then imagine trying to guess the flips that are the 5 Heads. If you pick a first flip at random, the chance of it correctly being a Head is $\frac{5}{10}$. If this is correct, the chance of getting the next choice right is $\frac{4}{9}$, since there are 9 coin flips left, 4 of which are Heads. This pattern continues for the third, fourth and fifth choice, and so overall the probability of correctly guessing the positions of the 5 Heads is

$$
\frac{5}{10} \times \frac{4}{9} \times \frac{3}{8} \times \frac{2}{7} \times \frac{1}{6} = \frac{1}{252}
$$

But the probability of guessing the positions of the 5 Heads is exactly the same as

$$
\frac{1}{\text{the number of possible positions for the 5 Heads}}
$$

so there are 252 different sequences, exactly the entry in Pascal's triangle.

This 'trick' generalises to a formula for the number of ways of arranging r Heads in n flips. Following the same reasoning as above, this turns out to be

$$
\frac{n}{r} \times \frac{n-1}{r-1} \times \ldots \times \frac{n-r+1}{1} = \frac{n!}{r!\,(n-r)!}
$$

where $r! = r(r-1)(r-2)\ldots1$ is known as 'r factorial'.

The quantity $\frac{n!}{r!(n-r)!}$ is traditionally denoted ^nC_r or $\binom{n}{r}$, so the rth entry in the nth row of Pascal's triangle is given by ^nC_r.

^nC_r is also the general formula for the number of different ways in which r objects can be chosen from n. If there are 10 people in a class, and a committee of 4 have to be chosen, how many possible different committees are there? The answer is $^{10}C_4 = \frac{10!}{4!\,(10-4)!} = \frac{10!}{4!6!} = \frac{10 \times 9 \times 8 \times 7}{4 \times 3 \times 2 \times 1} = 210$. The numbers always cancel nicely!

This analysis is used in Chapters 27 and 29.

Boys and girls in a family

This could bring in ideas of family planning, and so teacher discretion is advised.

For members of the class in families with the same number, n, of children, say $n = 1, 2, 3, 4, 5$, note how many girls there are in each family (in single sex classes, ask about siblings). Then construct the distribution (in a bar chart) and compare it with the shape of the nth row of Pascal's triangle, corresponding to the total number of children in the family. How good is the fit? After all, if the probability of a boy or a girl is 50–50, the distribution should be the same as that for coin flipping. If the fit is not all that good, why might that be?

There are a number of reasons why the fit may not be good. First, and most important, it is unlikely that enough families were considered to see whether the experimental distribution fits the theoretical distribution. Second, there might be a deficit of families which comprise only boys or only girls, since parents might keep on having children until they have at least one of each. Third, and least important, it is a curious fact that slightly more boys than girls are born. There were 695 233 live births in England and Wales in 2014, of which 338 461 were girls and 356 772 were boys [1], so $\frac{356\,772}{695\,233} = 51.3\%$ boys. Nobody really knows why this happens.

25.3 Further reading and resources

There is a good article on Pascal's triangle on Wikipedia [2].

25.4 References

1   Office for National Statistics. *Birth Characteristics in England and Wales, 2014* [Internet]. [cited 2015 Oct 23]. Available from: http://www.ons.gov.uk/ons/rel/vsob1/birth-characteristics-in-england-and-wales/2014/stb-birth-characteristics-2014.html

2   Wikipedia. *Pascal's triangle* [Internet]. [cited 2015 Oct 23]. Available from: https://en.wikipedia.org/wiki/Pascal%27s_triangle

Your risk is increased!

26.1 Summary

The media are full of warnings about activities that increase the risk of bad things happening to you. Here we show how to take those stories apart and see whether we should really worry.

26.2 Possible classroom activities

Bacon sandwiches

> This material deals with eating pork products, which is forbidden in some faiths. We also discuss bowel cancer. Teacher discretion is advised.

Under a headline 'Bacon butty cancer risk', a report in 2007 said that eating a bacon sandwich each day increased the risk of bowel cancer by 20%. In 2015 the World Health Organisation confirmed that processed meat (such as bacon and ham) increased the risk of cancer, with an estimated 18% increased risk of bowel cancer for an extra 50g of processed meat eaten per day [1].

Perhaps you would never eat a bacon sandwich because of your faith or being vegetarian, but, for people who would eat one, do you think this headline would affect their appetite for this food?

A 20% increase sounds a lot. But the crucial idea is that its significance depends on how common bowel cancer is.

> Around 1 in 20 people will get bowel cancer in their lifetime, so if someone eats a daily bacon sandwich, this chance is increased by 20%. Compare what would happen to 100 people who did not eat a daily bacon sandwich with what would happen to 100 people who did.

Out of 100 people who *did not* eat a daily bacon sandwich, we would expect 5 to get bowel cancer. Out of 100 people who *did* eat a daily bacon sandwich, we would expect 20% more, which means 6, to get bowel cancer.

The 20% increase is known as a **relative risk**, and it is a measure of change. The risk of actually getting cancer is known as the **absolute risk**. Because the relative risk is a 20% increase of a fairly small absolute risk, it actually means that we should expect just one extra person to get bowel cancer in every 100 who eat bacon sandwiches every day. We say that the absolute risk is increased by 1 percentage point, from 5% to 6%, shown graphically in icon arrays (Figure 26.1).

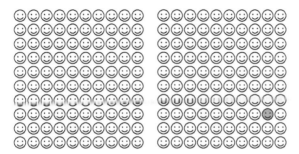

Figure 26.1 Icon arrays showing the additional risk of bowel cancer from bacon sandwiches: the 100 people on the left do not eat bacon, the 100 on the right eat bacon every day, and one extra case of bowel cancer is expected

The left-hand array shows the number of people we would expect to get bowel cancer (yellow) out of 100 people who *do not* eat a daily bacon sandwich. The right-hand array shows the number of people we would expect to get bowel cancer (yellow and blue) out of 100 people who *do* eat a daily bacon sandwich – the blue person is the additional person, whose cancer was caused by the daily bacon sandwich.

A substantial collection of teaching resources on this topic can be obtained from the Motivate programme website [2].

Rare risks

This material deals with drinking alcohol, which some faiths forbid or discourage. We also discuss mouth cancer. Teacher discretion is advised.

A newspaper headline said 'Two glasses of wine a day triples risk of mouth cancer' [5]. Assuming we believe the story, should people be worried? If mouth cancer affects 9 per 100 000 people per year, how many extra cases would we expect per 100 000 people drinking this much? How many people would have to drink this much to generate one extra case of mouth cancer each year?

Of 100 000 people drinking this much wine, instead of 9 there would be 27 getting mouth cancer each year, so that is an additional 18 people per 100 000 people getting mouth cancer each year. This means that $\frac{100\,000}{18} \approx 5500$ people would have to drink this much extra each day, to generate one extra case of mouth cancer per year.

(Also, these are large 250 ml glasses, so this amount of wine contains around 6 units of alcohol, three times the recommended UK daily limit and a rather large consumption.)

26.3 Further reading and resources

The Understanding Uncertainty website contains an animation showing 2845 ways to display risks [6]. Students could also be asked to keep a lookout for similar stories of exaggerated risks in the news – there is even a website that lists all of the cancer-causing/cancer-curing claims that tabloid newspapers have made [7].

26.4 References

1   Gallagher J. *Processed meats do cause cancer – WHO* [Internet]. BBC News [cited 2015 Nov 4]. Available from: http://www.bbc.com/news/health-34615621

2   Spiegelhalter D. *"Eating bacon sandwiches is bad for you!" Evaluating Risk* [Internet]. [cited 2015 Oct 25]. Available from: https://motivate.maths.org/content/MathsHealth/Risk

3   xkcd. *Increased Risk* [Internet]. [cited 2015 Oct 25]. Available from: http://www.xkcd.com/1252/

4   Explain xkcd. *Increased Risk* [Internet]. [cited 2015 Oct 25]. Available from: http://www.explainxkcd.com/wiki/index.php/1252:_Increased_Risk

5   Manning S. *Two glasses of wine a day triples risk of mouth cancer* [Internet]. The Independent [cited 2015 Oct 25]. Available from: http://www.independent.co.uk/life-style/health-and-families/health-news/two-glasses-of-wine-a-day-triples-risk-of-mouth-cancer-6432609.html

6   Understanding Uncertainty. *2845 ways to spin the Risk* [Internet]. [cited 2015 Oct 25]. Available from: http://understandinguncertainty.org/node/233

7   *Kill or cure?* [Internet]. [cited 2016 Feb 4]. Available from: http://kill-or-cure.herokuapp.com/

Chapter 27

It's a lottery

27.1 Summary

Strictly speaking, lotteries are a mild form of gambling, which is forbidden or discouraged in many faiths. Teacher discretion is advised.

Lotteries are a good opportunity to illustrate many issues in probability, and here we look at calculating the chances of winning, patterns in numbers, and probability myths. These explorations should, with luck, discourage students from playing the lottery.

27.2 Possible classroom activities

A simple '3/6' lottery

Experiment: Label six counters 1 to 6 and put them in an opaque bag. Each student should choose and write down four imaginary lottery tickets, bearing three different numbers between 1 and 6, for instance, 3-4-6. Draw three counters out of the bag in turn, taking time between each draw to see who is still in the running for a 'prize' if they match all three numbers drawn. This is known as a '3/6' lottery as the aim is to choose three numbers correctly from six. Repeat the experiment five times so each student has had 20 attempts to win. Count up the number of 'wins' for each student.

Who did best? What was the overall proportion of wins in the class?

Simulation: Try the simulation provided by NRICH [1]. This allows you to pick a 'ticket', randomly generate multiple draws for different types of lottery, then label the matching balls and sort the results. In 50 draws in a 3/6 lottery, what proportion of 'wins' is recorded?

Theoretical: For this lottery, write out all the different tickets that can be chosen (remember that 3-4-6 is the same ticket as 6-4-3). How many different tickets are there?

The first number on the ticket can be any of 6, the second any of the remaining 5, the third any of the remaining 4, giving $6 \times 5 \times 4 = 120$ different tickets, if the order of the numbers were important. Since the order is not important, and any three different numbers can be written in six different orders, there are actually only 20 different tickets.

This is the same as the number of ways 3 objects can be chosen from 6, which is $^6C_3 = \frac{6!}{3!(6-3)!} = \frac{6!}{3!3!} = \frac{6 \times 5 \times 4}{3 \times 2 \times 1} = 20$. See Chapter 25: *Heads or Tails, boy or girl?* for the derivation of this, and the relation to Pascal's triangle.

The chance of winning is equal to $\dfrac{1}{\text{the number of different tickets}}$. If there are N different tickets, one of them must be the winner. Since each is equally likely to be the winner, the chance of picking the winning combination is $\frac{1}{N}$, so the chance of winning is $\frac{1}{20}$.

We can also obtain this directly using a probability argument. Imagine picking the counters out of the bag one at a time. The chance that the first matches one of your numbers is $\frac{3}{6}$. If this happens, there are 5 counters left, and the chance that the second matches one of your remaining numbers is $\frac{2}{5}$. If this happens, then the chance that the remaining counter matches your final number is $\frac{1}{4}$. Multiplying these together, since all the events must happen for you to win, we obtain $\frac{3}{6} \times \frac{2}{5} \times \frac{1}{4} = \frac{1}{20}$ for the probability of a win. This is by far the simplest way of obtaining the probability of a win, and neatly avoids any discussion of permutations and combinations.

The UK Lotto game

In the current UK Lotto game, you pay £2 for a ticket, and mark six different numbers between 1 and 59. Then on Wednesday and Saturday 59 balls bounce around in a plastic container and six are drawn (plus a bonus ball). You share the jackpot if your ticket matches the six numbers drawn. In October 2015 the jackpot fund ranged between £3 million and £12 million. Is it worth buying a ticket?

Around 48% of ticket sales is paid out in prizes. There is a complex prize system, but most of the prize money (93%) goes either on the jackpot (30%) or on a large number of small prizes (63%) from matching two or three balls. Why has this prize structure been chosen? Presumably it is based on a belief that a tiny chance of a life-changing win is the reason people buy lottery tickets, but unless players occasionally win something they will become discouraged and stop buying tickets.

The UK Lotto is a '6/59' lottery as 6 balls are chosen from 59, although before October 2015 it was a 6/49 lottery, as are many major lotteries. If you buy a single ticket in a 6/59 lottery, what is the chance of winning the jackpot?

Like the simple '3/6' lottery, there are two ways to calculate this. We start with the easier, 'sequential' method. Imagine the balls being drawn one by one. The chance that the first ball drawn matches one of your numbers is 6/59. If this happens, the chance that the second ball drawn matches one of your remaining numbers is 5/58. And so on until the chance that the final ball drawn matches your final number is 1/54. Multiplying these together, since all the events must happen for you to win the jackpot, we obtain that the probability of winning the jackpot is $\frac{6}{59} \times \frac{5}{58} \times \frac{4}{57} \times \frac{3}{56} \times \frac{2}{55} \times \frac{1}{54} = \frac{1}{45\,057\,474}$.

This is a very low chance indeed. There are far fewer than 45 million tickets sold each week, and so jackpot winners are now rare. How many jackpot winners have there been in the last 10 draws? (Information is available on the National Lottery website under *Results* [2].)

The alternative method of obtaining the probability is to calculate the total number of possible tickets. This is the same as the number of ways 6 objects can be chosen from 59, which is $^{59}C_6 = \frac{59!}{6!(59-3!)} = \frac{59!}{6!\,56!} = 45\,057\,474$, and this, reassuringly, is the inverse of the equation above.

If you buy a single ticket in a 6/49 lottery, as the UK Lotto was between 1994 and 2015, what is the chance of winning the jackpot? The sequential method leads to $\frac{6}{49} \times \frac{5}{48} \times \frac{4}{47} \times \frac{3}{46} \times \frac{2}{45} \times \frac{1}{44} = \frac{1}{13\,983\,816}$.

> Other questions that we can ask include:
>
> In what proportion of winning combinations are there two adjacent numbers? (Winning combinations can be obtained from the National Lottery website.)
>
> What is the chance that two adjacent numbers will come up on a winning ticket?

We can use past winning tickets to count the proportion of winners with adjacent numbers, and so obtain an estimate of the probability of adjacent numbers. Getting the theoretical probability is more tricky – who would like to count how many of the 45 057 474 possible combinations have adjacent numbers? One way to approach this is to get a rough probability that there are *no* adjacent numbers (as in working out the chance of birthday coincidences, Chapter 31). The first number drawn can be anything. The second drawn cannot be adjacent, and so must be at least 2 away from the first one, which happens with probability $\frac{56}{58}$ since there are 58 numbers left but two are ruled out as being adjacent to the first. Having drawn the second, non-adjacent ball, up to four numbers are now ruled out, so the chance that the third number is not adjacent to the previous two is around $\frac{53}{57}$. And so on, giving an overall probability of no adjacencies of around $\frac{56}{58} \times \frac{53}{57} \times \frac{50}{56} \times \frac{47}{55} \times \frac{44}{54} = 0.56$. Thus the chance of adjacent numbers is around 44%, or nearly $\frac{1}{2}$. Note that this ignores all sorts of issues of overlaps and boundaries!

Lottery myths

> 'Many winning lottery tickets have adjacent numbers, so it makes sense to pick a ticket with adjacent numbers.'
>
> 'Most winning combinations are a mixture of odd and even numbers, so you should choose a mixture of odd and even numbers.'
>
> Is this advice sensible?

Claims like these have been made on various websites. They cannot make sense, as every ticket has the same chance of winning, so no choice can increase your chances of getting a match. But it is not easy to say exactly what is wrong with such arguments.

For the second claim, the basic error is to confuse the following:

* the probability of a ticket having a mixture of odd and even numbers, given it has won (which is very high)

- the probability of a ticket winning, given it has a mixture of odd and even numbers (which is very low, and the same for every ticket).

Confusing these two different conditional probabilities is known as the 'prosecutor's fallacy'. One way to view the problem is to realise that although the set of 'tickets with a mixture of odd and even numbers' is likely to include the winner, this is only because there are so many such tickets in the set – the chance of your *particular* ticket winning is still the same as all the others.

Does it make sense to say a number is 'due'?

No. How are those bouncing balls supposed to have a memory? Of course some numbers are bound to come up more times than others, by chance alone (Chapter 21). Indeed, once a number is in the lead, it tends to stay there – '38' was in the lead for years, while '13' lagged at the bottom. But this means nothing; it's just chance.

See the Understanding Uncertainty website for animations of lottery histories [3].

Lottery analogies

The chance of winning the jackpot in a 6/59 lottery is 1 in 45 057 474. How many times would you have to flip a coin, to have roughly the same chance of getting Heads every time?

The chance of flipping 25 Heads in a row is 1 in 2^{25} = 33 554 432, so winning the jackpot is less likely than flipping 25 Heads in a row.

How many times would you have to throw a die, to have roughly a 1 in 45 057 474 chance of getting a 'six' every time?

The chance of throwing 10 sixes in a row is 1 in 6^{10} = 60 466 176. So winning the jackpot is slightly more likely than throwing 10 sixes in a row.

Try the following analogy for the Lotto jackpot.

Take a bath and fill it with rice. Then take one grain of rice, paint it gold and bury it somewhere in the bath. Blindfold your friend, take their £2 and invite them to plunge their hand in the bath and pick up one grain of rice. If they pick the golden grain, they win £5 million. If the chance of them winning is the same as winning the lottery (1 in 45 057 474), how deep must the bath be filled with rice? (Assume there are around 60 grains of rice in a cm³ and that the bath has a rectangular cross-section, 70 cm wide by 170 cm long.)

We need 45 057 474 grains in the bath, including the golden one. There are 60 grains in a cm³, so we need a volume of $\frac{45\,057\,474}{60}$ = 750 958 cm³. The cross-section of the bath is 70 × 170 = 11 900 cm², so the bath must be $\frac{750\,958}{11\,900}$ = 63 cm deep in rice (around 2 feet). This will make most baths overflow! Do not try this at home.

27.3 Further reading and resources

See Wikipedia [4] and the National Lottery website [5] for extensive details on the lottery, including spreadsheets of past draws, and *Teaching mathematics through the national lottery* by Burghes and Galbraith [6].

Plus magazine has an excellent article by John Haigh [7] on choosing numbers in the lottery – as you can't change your chances of winning, the only thing is to try and choose combinations that others don't choose, so that you don't have to share the prizes.

The website contains a spreadsheet that calculates chances for different types of lottery.

27.4 References

1   NRICH. *Mathsland National Lottery* [Internet]. [cited 2015 Oct 24]. Available from: http://nrich.maths.org/7238

2   The National Lottery. *Lotto draw history* [Internet]. [cited 2015 Oct 24]. Available from: https://www.national-lottery.co.uk/results/lotto/draw-history?icid=mdr%3Alo%3Atx

3   Understanding Uncertainty. *Lottery expectations* [Internet]. [cited 2015 Oct 24]. Available from: http://understandinguncertainty.org/node/40

4   Wikipedia. *National Lottery (United Kingdom)* [Internet]. [cited 2015 Oct 24]. Available from: https://en.wikipedia.org/wiki/National_Lottery_(United_Kingdom)

5   The National Lottery [Internet]. [cited 2015 Oct 24]. Available from: https://www.national-lottery.co.uk/

6   Burghes D, Galbraith P. *Teaching Mathematics through National Lotteries* [Internet]. International Journal for Mathematics Teaching and Learning. 2000 Feb 22 [cited 2015 Oct 24]. Available from: http://www.cimt.plymouth.ac.uk/journal/ijnatlot.pdf

7   Haigh J. *The UK National Lottery – a guide for beginners* [Internet]. [cited 2015 Oct 24]. Available from: https://plus.maths.org/content/os/issue29/features/haigh/index

Chapter

28

Switch or stick?

28.1 Summary

The classic *Monty Hall* problem is based on an old American TV game show. The 'standard' description of the game is as follows: the presenter, Monty Hall, presents a contestant with three identical doors, behind two of which stands a goat, and the other hides a car. The contestant is invited to choose a door, but it is not opened. Monty Hall, who knows where the car is, then opens one of the two doors that has not been chosen, revealing a goat (if Monty has a choice of two doors to open, we assume he picks one at random). The contestant is then offered two options: switch doors or stick to their original choice?

It may appear that, with now just two doors left, the chances that the car is behind the original chosen door is $\frac{1}{2}$, and so there is no point in switching. However, rather unintuitively, the probability that the car is behind the 'other' door is $\frac{2}{3}$, and so it is always better to switch. This solution has been subject to considerable dispute.

Although the problem is very familiar within the probability community, it is generally unknown to students. It can be irritating if it comes over as just an unintuitive and baffling result, and so it is best thought of as an example of testing alternative strategies by considering the question: 'what would I expect to happen if I kept on using this strategy many times?' We consider the standard version, and suggest multiple ways of trying to explain the counter-intuitive result.

28.2 Possible classroom activities

Students are put into pairs, with each pair having 3 paper cups and a number of small sweets (or a single small object if it is not appropriate to consume sweets). Each student alternates between being Player A (Monty Hall) and Player B (contestant). Player B closes their eyes while Player A places one sweet/object under one of the cups and *remembers which one*. Player B then opens their eyes, and points at one of the three cups. Player A lifts one of the *other* cups revealing it to be empty – this is always possible as at least one of the cups not chosen is bound to be empty, and if there are two empty cups, one should be chosen at random. Should Player B now switch cups, or stick with their original choice? Having decided, Player B either sticks or switches, lifts the chosen cup and records whether they got the sweet or not.

Students should keep a tally of their results (Table 28.1), noting whether the decision was to stick or switch.

Player hiding	Player guessing	Switch or stick?	Win or lose?
X	Y	switch	won
Y	X	stick	lost
...	...	...	...

Table 28.1 Monty Hall results

Stop after a couple of games (when students understand the game) and discuss what they feel. Then carry on and tally up at the end.

What strategy do students choose? Do they think it makes a difference? Our intuition is usually that, after the empty cup is lifted, there are now two options, so the chances are 50–50 and it makes no difference whether you stick or switch. However, after playing for a while and tallying up, it should be clear that switching means you win roughly twice as often as sticking. Why? Can anyone provide an explanation?

Note: the game needs to be described carefully. If Player B starts off pointing to the cup with a sweet, Player A has two options as to which cup they lift, and they should choose randomly. Player A must also remember where the sweet is, so they always lift an empty cup.

There are a huge number of variations of the game, each with different properties. Some pairs could repeat the game with more cups, up to ten if possible, but still with only one sweet. After Player B points at a cup, Player A then lifts all but one of the empty cups, so lifts eight empty cups, leaving only two to choose from. It should then be far more intuitive that it is better to switch.

There are multiple ways of trying to explain why it is better to switch, some of which should come up in discussion.

Call the cup that Player B points at initially Cup 1. There is probability $\frac{1}{3}$ of Player B picking the right cup to start with, that is, Cup 1 covers the sweet. When Player A lifts an empty cup (which we call Cup 2), this provides no additional information, since Player A will always reveal an empty cup, regardless of whether Cup 1 covered the sweet or not. The initial probability of $\frac{1}{3}$ is therefore unchanged, which means the probability that Cup 3 covers the sweet is $\frac{2}{3}$.

Another approach, which we prefer, is to see that if Player B adopts the strategy of switching, they will only lose if Cup 1 actually covered the sweet. Since this happens with probability $\frac{1}{3}$, a switching strategy will only lose $\frac{1}{3}$ of the time.

An expected frequency tree (Figure 28.1) can be constructed to show the expected number of wins over, say, 30 repeats of the game using the different strategies. We can assume that Cup 1 is expected to truly cover the sweet in 10 out of 30 cases, and the labelling of Cup 2 and Cup 3 is arbitrary.

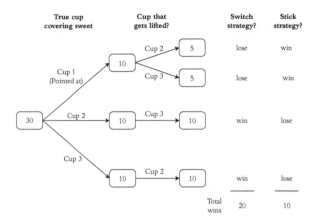

True cup covering sweet		Cup that gets lifted?		Switch strategy?	Stick strategy?
		Cup 2 → 5		lose	win
	10				
Cup 1 (Pointed at)		Cup 3 → 5		lose	win
30	Cup 2 → 10	Cup 3 → 10		win	lose
	Cup 3				
	10	Cup 2 → 10		win	lose
			Total wins	20	10

Figure 28.1 Expected frequency tree for the Monty Hall problem

Falk has come up with a nice analogy which makes the solution even more baffling [1]:

> Tom, Dick and Harry are all equally suspected of a crime that one of them has definitely committed. Tom is picked at random and brought in for questioning, but just before he arrives at the police station, Dick walks in with a photograph that provides himself with an alibi. Should the police now switch their attention to Harry rather than Tom?

It is true, but rather unintuitive, that they should switch their attention. Again, it is helpful to extend the scenario to assuming there are 100 suspects, and after picking Tom at random for questioning, 98 of the remaining suspects come up with an alibi – that is, everyone else except Harry. It then seems more reasonable to switch attention to Harry.

28.3 Further reading and resources

There is extensive discussion of the problem on many websites and YouTube. Wikipedia [2] provides multiple explanations and a good history of the controversy surrounding the dilemma, and shows a solution using Bayes theorem in full probability notation.

Puza [3] provides an algebraic analysis of many different versions of the game.

28.4 References

1 Falk R. *Monty's dilemma with no formulas.* Teaching Statistics. 2014 Jun 1;36(2):58–61.

2 Wikipedia. *Monty Hall problem* [Internet]. [cited 2015 Oct 17]. Available from: https://en.wikipedia.org/wiki/Monty_Hall_problem

3 Puza BD, Pitt DGW, O'Neill TJ. *The Monty Hall Three Doors Problem.* Teaching Statistics. 2005 Feb 1;27(1):11–5.

It's not fair!

29.1 Summary

The idea of 'fairness' arises in early childhood. In games of chance between two people, it means that the two players have equal chances of winning. In more complex situations, rewards or losses will vary according to outcomes, and the crucial concept is the 'expected return', where expectation is the average of the consequences of different outcomes, weighted by the probability of their occurrence. In notation, if there are n possible outcomes with probabilities $p_1, \ldots, p_n$, and the values of these outcomes are $v_1, \ldots, v_n$, the overall expected value is $E = p_1 v_1 + \ldots + p_n v_n$.

Expected return trees are a natural extension of probability trees in which consequences are attached to final outcomes. Expected frequency trees can be similarly extended to illustrate long-term expected gains and losses. We provide a variety of games of differing complexity to illustrate these ideas and then show that insurance is just one more form of a (unfair) game. In this section, we are assuming fairly small gains and losses so that simple expectation is the appropriate measure for comparison. More complex situations with larger gains and losses, where psychological attitudes to risk become important, are dealt with in Chapter 30: *Take a risk?*

Some of this section necessarily deals with forms of gambling, if only for points, and so needs to be handled with care. The idea is to discourage gambling by showing that the house always wins in the end.

29.2 Possible classroom activities

Is this fair? Counter game

There are 2 red and 4 blue counters in a bag. Sara takes out two of them at random. If they are the same colour, Sara wins. If they are different colours, Alan wins. Is this a fair game? In a two-person game like this one, 'fair' just means that each person has an equal chance of winning.

The tree below (Figure 29.1) shows the chances of each person winning.

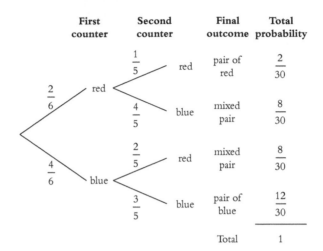

First counter	Second counter	Final outcome	Total probability
$\frac{2}{6}$ red	$\frac{1}{5}$ red	pair of red	$\frac{2}{30}$
	$\frac{4}{5}$ blue	mixed pair	$\frac{8}{30}$
$\frac{4}{6}$ blue	$\frac{2}{5}$ red	mixed pair	$\frac{8}{30}$
	$\frac{3}{5}$ blue	pair of blue	$\frac{12}{30}$
		Total	1

Figure 29.1 Probability tree for counter game

The probability that Sara wins is $\frac{2}{30} + \frac{12}{30} = \frac{14}{30}$, which is less than $\frac{1}{2}$, and so the game is not fair – it favours Alan.

Suppose there are r red counters and b blue counters in the bag. Show that the game is 'fair' if $(r - b)^2 = r + b$. Find some examples of r and b that obey this equation. (Hint: their sum must be a perfect square.)

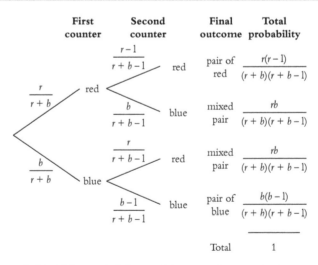

First counter	Second counter	Final outcome	Total probability
$\frac{r}{r+b}$ red	$\frac{r-1}{r+b-1}$ red	pair of red	$\frac{r(r-1)}{(r+b)(r+b-1)}$
	$\frac{b}{r+b-1}$ blue	mixed pair	$\frac{rb}{(r+b)(r+b-1)}$
$\frac{b}{r+b}$ blue	$\frac{r}{r+b-1}$ red	mixed pair	$\frac{rb}{(r+b)(r+b-1)}$
	$\frac{b-1}{r+b-1}$ blue	pair of blue	$\frac{b(b-1)}{(r+h)(r+b-1)}$
		Total	1

Figure 29.2 Probability tree for the general case of the counter game

The game is fair if the probabilities of the two players winning are the same. From the probability tree (Figure 29.2) we see this holds if $r(r - 1) + b(b - 1) = 2rb$, which can be rearranged to $(r - b)^2 = r + b$.

The right-hand side, $r + b$, must be a perfect square, since it equals $(r - b)^2$. So suppose there are $r + b = 4$ counters altogether, the difference between r and b must be $\sqrt{4} = 2$, so $r = 1$ and $b = 3$ or vice versa. Other solutions for (r, b) are $(3, 6)$, $(6, 10)$, $(10, 15)$ and so on: the pattern should be clear.

Hence show that the game is 'fair' if r and b are equal to $\frac{x(x+1)}{2}$ and $\frac{x(x-1)}{2}$ (either way round) for integers $x = 1, 2, 3, 4, \ldots$ [Note: these give the triangle numbers, which is not surprising since $T_n - T_{n-1} = n$, and $T_n + T_{n-1} = n^2$, where T_n is the nth triangle number and T_{n-1} the $(n-1)$th.]

Let $x = r - b$. Then there are a total of $r + b = x^2$ counters. Solving these simultaneous equations for r and b gives the solution.

This is adapted from NRICH's *In a Box* [1].

Is this fair? Coin flipping

You pay 2 points to enter a game in which a coin is flipped. If the coin comes up Heads, you win nothing. If it comes up Tails, you win 3 points. Is this 'fair'? How much would you expect to win/lose if you played this game 10 times? Would you want to play this game?

The idea is to include rewards on the probability tree and calculate the expected return (Figure 29.3).

The final column is 'probability × return', which when added over the branches gives the expected return. But the initial stake needs to be subtracted in order to give the overall expected gain or loss. In this case we expect, on average, to lose 0.5 points per game. This is perhaps better expressed as the expected return when playing a number of games (Figure 29.4).

Figure 29.4 shows that over 10 games, we would expect to lose 5 points. We would not recommend playing this game!

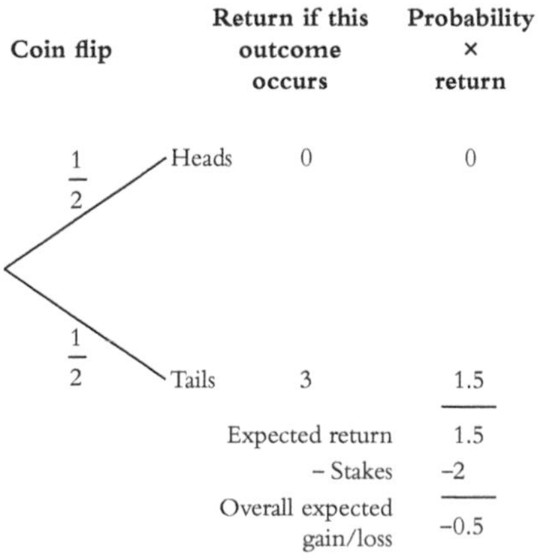

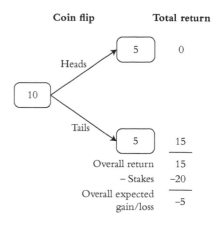

Figure 29.3 Expected return tree for coin flipping game

Figure 29.4 Expected frequency tree when playing the coin flipping game 10 times

How much should Mr Biggins charge?

A teacher, Mr Biggins, invents the following game for the school fete. The game board is shown in Figure 29.5. A counter starts on the square marked 'Start'. The player flips a coin four times, moving horizontally on a Head, and vertically on a Tail. The player wins if they end on the square marked 'Win'.

The prize for winning is 40p. What is the chance of winning? How much should Mr Biggins charge if:

- he expects to make a profit?
- he expects to make £10 profit from 100 players?

This is adapted from Dan Walker's slides [2].

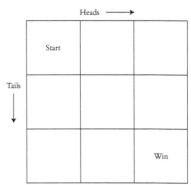

Figure 29.5 Mr Biggins's game board

To win, a player must move two squares vertically and two horizontally, so needs 2 Heads and 2 Tails in four coin flips. There are 16 possible sequences for four coins to land. In six of these there are 2 Heads and 2 Tails, so the chance of winning is $\frac{3}{8}$ (see Chapter 25: *Heads or Tails, boy or girl?*). The expected return when someone plays is therefore $\frac{3}{8} \times 40 = 15$p, so Mr Biggins must charge at least 15p to expect to make any profit. To make £10 from 100 players, he needs to expect to make 10p per game, so he should charge 25p.

Which is the better bet?

Which of the following is the better bet, if both games cost £1 to play?

Gamble 1: Getting 2 Heads and 2 Tails on four coins wins you £3.

Gamble 2: You win £2 for every '6' that appears when three standard dice are rolled.

Gamble 1: There are 16 possible sequences for four coins to land. In six of these there are 2 Heads and 2 Tails, so the chance of winning Gamble 1 is $\frac{3}{8}$ (see Chapter 25: *Heads or Tails, boy or girl?*) and the expected return is $\frac{3}{8} \times 3 = \frac{9}{8}$.

Gamble 2: We need the expected number of '6's if we throw three dice. There is a very long and difficult way to do this: work out the probabilities of getting zero, one, two or three '6's, and so find the expected number. Alternatively, and much simpler, we know by symmetry that the expected number of '6's, if we throw a die six times, is 1, so the expected number of '6's in three throws must be $\frac{1}{2}$. So the expected return from Gamble 2 is $\frac{1}{2} \times 2 = 1$.

Therefore Gamble 1 is preferable, and the expected gain per gamble is 'expected return − stake' $= \frac{9}{8} - 1 = \frac{1}{8}$.

See NRICH's *The Better Bet* [3], which features an animation to explore these experimental outcomes.

Why the house always wins in the end

The game of European roulette is played by placing chips on areas of a table depending on which number you predict will come up when a ball is spun on the roulette wheel. The wheel has 37 slots, numbered 1 to 36 plus a '0', and each number 1 to 36 also has a colour, red or black. If '0' comes up, you lose all your bets.

Some of the possible bets are:

- If you put one chip on 'red' and the number that comes up is red, you get two chips back, and similarly for 'black'.

- If you put one chip on a specific number, say '1', and it comes up, you get 36 chips back.

What is my overall expected gain/loss if I put one chip on 'red'?

What is my overall expected gain/loss if I put one chip on '1'?

Do you think you can 'beat the house' in roulette?

This can be solved by extending a probability tree to include consequences (Figure 29.6).

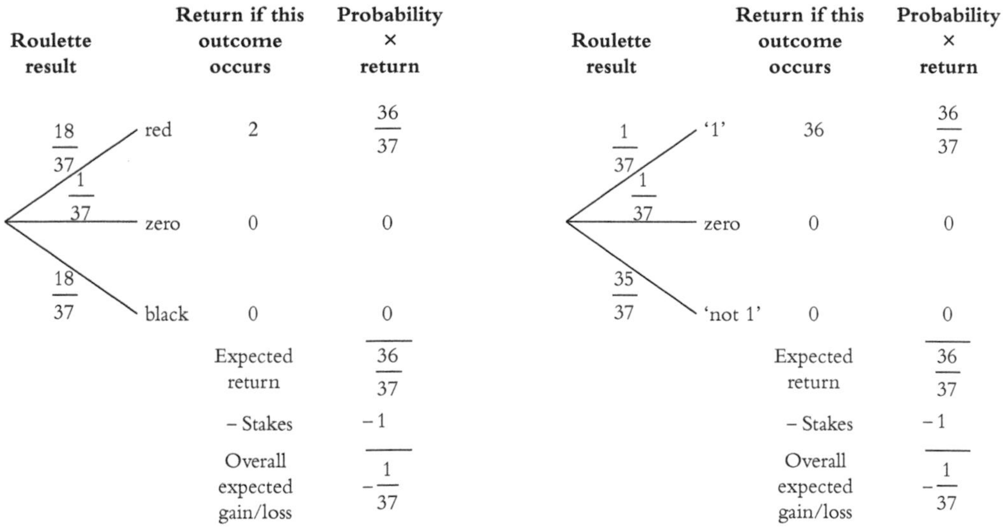

Figure 29.6 Roulette expected return trees for betting on 'red' (left) and betting on '1' (right), assuming our stake is 1 unit

This could also have been carried out using an expected frequency tree based on 37 bets. In each case my overall expected loss per game is $\frac{1}{37}$ = 2.7% of my stake.

This is known as the 'house advantage' – the casino expects, on average, to make 2.7% of all stakes made. In the short term a gambler may be lucky (there is no skill whatsoever in roulette), but over time the house always wins. You have been warned.

Under Over

Here is an old fairground game: two dice will be thrown, and you guess whether you think the total will be 'under 7', 'exactly 7', or 'over 7'. You pay 1 point to play. Your winnings are:

- if you guessed 'under 7' and you are right, you get 2 points back
- if you guessed 'exactly 7' and you are right, you get 5 points back
- if you guessed 'over 7' and you are right, you get 2 points back.

Play the game 10 times, and see whether the 'banker' tends to win. Show that your expected return is the same, whatever you bet on. What is the house advantage?

There are 36 different possibilities for the two dice. In six of these the sum is exactly 7, so the chance of getting 'exactly 7' is $\frac{6}{36}$. The chance of getting 'under 7' is $\frac{15}{36} = \frac{5}{12}$, and, by symmetry, this is the same as the chance of 'over 7'. The expected return tree is shown in Figure 29.7, assuming two different bets: that the result will be 'under 7' and that the result will be 'exactly 7'.

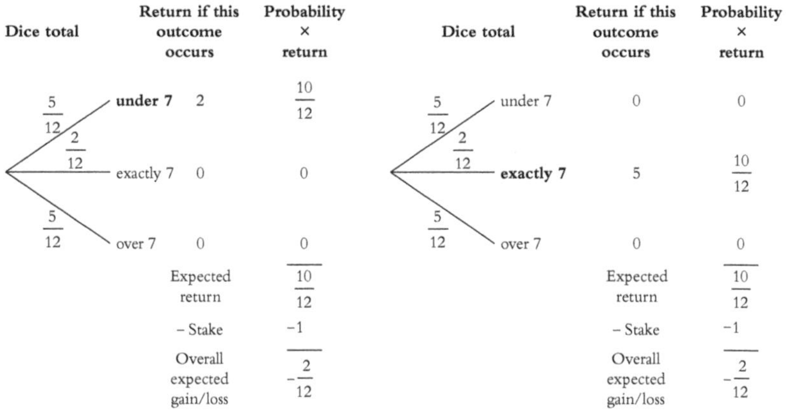

Dice total	Return if this outcome occurs	Probability × return		Dice total	Return if this outcome occurs	Probability × return
$\frac{5}{12}$ under 7	2	$\frac{10}{12}$		$\frac{5}{12}$ under 7	0	0
$\frac{2}{12}$ exactly 7	0	0		$\frac{2}{12}$ **exactly 7**	5	$\frac{10}{12}$
$\frac{5}{12}$ over 7	0	0		$\frac{5}{12}$ over 7	0	0
	Expected return	$\frac{10}{12}$			Expected return	$\frac{10}{12}$
	– Stake	–1			– Stake	–1
	Overall expected gain/loss	$-\frac{2}{12}$			Overall expected gain/loss	$-\frac{2}{12}$

Figure 29.7 Expected return tree for *Under Over*

So whatever the bet, on average you can expect to lose $\frac{2}{12} = \frac{1}{6}$ of your stake at each game.

The house advantage is $\frac{1}{6}$ = 17%, so on average they take 17% of the stakes placed at each game.

Insurance is a 'game'

Tom buys a smartphone for £100. The shop offers annual insurance that will replace the phone if it is lost, stolen or damaged. The insurance costs £10. Tom reckons there is around a 1 in 20 chance of his phone getting lost, stolen or damaged each year, and if he doesn't have insurance, he will need to buy a replacement phone. What is his overall expected loss over the year, if he:

- buys insurance?
- does not buy insurance?

If he does buy insurance, he knows it will cost him £10 for the year. The expected return tree (Figure 29.8) analyses the cost if he does not buy insurance.

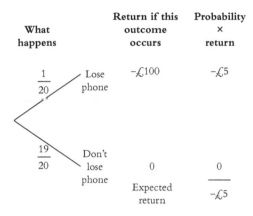

Figure 29.8 Expected return tree for insuring a mobile phone

So his expected loss when he does not buy insurance is £5, compared to £10 if he does buy insurance, and by this calculation it seems he should not buy insurance – the 'game' is against him.

> The insurance company has 1000 customers like Tom. How many do they expect to claim? What is their expected profit per customer over the year?

The expected frequency tree (Figure 29.9) shows what the insurance company expects from 1000 customers. 50 (or 5%) are expected to lose their phones, and these cost the insurance company £5000, but this is easily outweighed by the 950 who do not lose their phones and yet have paid their premiums.

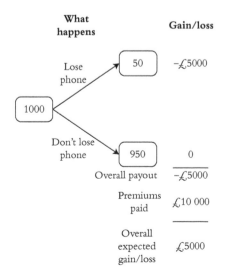

Figure 29.9 Expected return tree for insurance company

Because they cannot face a major loss – maybe Tom cannot afford another £100 if he loses his phone. Individuals do not have the opportunities for the gains and losses to cancel out, but an insurance company with many customers can allow the premiums paid to outweigh the losses. Insurance rests on the idea of sharing risk (but it only works if the insurance company uses probability theory correctly).

See Chapter 30: *Take a risk?* for a discussion about why people often choose options with a negative expected gain.

Adapted from the NRICH article *To Insure or Not to Insure* [4].

29.3 Further reading and resources

More material on the expected return from games can be found in Dan Walker's slides [2], while there are many articles on teaching probability with roulette [5, 6] and *Under Over* [7, 8].

29.4 References

1   NRICH. *In a Box* [Internet]. [cited 2015 Oct 26]. Available from: http://nrich.maths.org/919

2   Walker D. *Probability* [Internet]. TES Resources. [cited 2015 Oct 24]. Available from: https://www.tes.com/teaching-resource/probability-6321017

3   NRICH. *The Better Bet* [Internet]. [cited 2015 Oct 26]. Available from: http://nrich.maths.org/4334

4   NRICH. *To Insure or Not to Insure* [Internet]. [cited 2015 Oct 26]. Available from: http://nrich.maths.org/9598

5   Marshall JB. *Probability with Roulette*. Teaching Statistics. 2007 Aug 1;29(3):74–9.

6   Barr G, Scott L. *Teaching Statistical Principles with a Roulette Simulation* [Internet]. [cited 2015 Oct 26]. Available from: http://epublications.bond.edu.au/cgi/viewcontent.cgi?article=1137&context=ejsie

7   McPherson SH. *Unders and Overs: Using a Dice Game to Illustrate Basic Probability Concepts*. Teaching Statistics. 2015 Mar 1;37(1):18–22.

8   Wikipedia. *Under Over* [Internet]. [cited 2015 Oct 26]. Available from: https://en.wikipedia.org/wiki/Under_Over

Take a risk?

Chapter

30

30.1 Summary

This section covers the basic question: which would you prefer –
certainty or taking a chance? Activities are therefore based around
personal judgements, and there are no right answers. The crucial
comparison is between a *certain* gain or loss, and a 'risk' which can be
summarised by an *expected* gain or loss. People do not necessarily take the
option with the highest expectation – we have already seen in Chapter
29: *It's not fair!* that people may buy insurance even when their expected
costs are more than if they do not bother. Here we shall see that responses
can also be influenced by the 'framing' of the question.

The activities in this section could be considered a form of gambling,
although technically they come under the heading of 'behavioural
economics', so it should be made clear that these are completely fictional
situations designed to explore attitudes to taking risks and calculating
expectations.

Some of this material is concerned with monetary 'gambles',
although they are deliberately stylised as they are intended to
illustrate attitudes to risk rather than realistic gambling situations. We
have presented these tasks as if they were part of an imaginary TV
game show. Teacher discretion is advised.

30.2 Possible classroom activities

Play safe or take a chance?

This is a simple choice experiment concerning reasonably small gains.
Students are asked to imagine a TV game show in which they are offered
two choices, A or B, and have to indicate which they would prefer by a
show of hands.

Which would you prefer?

A   Receive £5 for certain.

B   Flip a coin – if it comes up Heads you get £X, if Tails you get nothing.

Start with X = 5. Clearly everyone should say they prefer A, the certain gain. Then try X = 10. The expected gains for A and B are now the same – each is £5. It is likely that most students will prefer the certain gain, but some risk-seekers might choose B. A real coin could be flipped to add some drama, even though this is all totally fictitious.

Then try X = 15. What is the expected gain from B? It is £7.50, which is more than the certain £5 from option A, but some cautious people might still prefer option A. Try X = 20, 50, 100, 1000. See how the proportion preferring option B increases (and if risk attitudes depend on gender!). If someone chooses the safe option when a risky alternative option has a higher expected gain, they are being 'risk-averse'.

Framing

The following classic question illustrates how the way in which a question is 'framed' can influence the responses.

Separate the class down the middle. Get one half to turn their backs, and put up the following choice, say on a whiteboard, asking them to silently write down whether they prefer A or B.

600 people are affected by a deadly disease and you have to choose between two treatments, A or B.

A   200 lives will be saved.

B   A 33% chance of saving all 600 people, 66% probability of saving no one.

Then get the other half to make a choice between the options:

A   400 people will die for certain.

B   A 33% chance that no people will die, 66% probability that all 600 will die.

When both sets of choices are revealed, it should become clear that not only are they identical, but the expected number of deaths (400) is the same for all four of the options presented. Compare what the two groups chose. If they follow previous experiments, in the first choice students will generally tend to choose option A, while in the second choice they will generally choose option B.

When choices are framed in a 'positive' way (lives saved), people tend to be risk-averse and choose the safe option. In contrast, when choices are framed in a 'negative' way (lives lost), people tend to be risk-seeking and choose the risky option. Psychologists have called this the 'framing effect' [1].

What would you prefer? Gains and losses

The right-hand side of the class turn their backs, while the left-hand side silently write down their answers to the following question.

In a TV game show, you are given £5, and then have to choose between

A   £10 more

B   50% chance of £20 more, 50% chance of £0 more.

Which would you choose, A or B?

Now the left-hand side of the class turn their backs, while the right-hand side silently write down their answers to the following question.

In a TV game show, you are given £25, and then have to choose between

A   returning £10

B   50% chance of returning £0, 50% chance of returning £20.

Which would you choose, A or B?

When both 'games' are revealed to the whole class, it should become apparent that they are *exactly* the same decision, and in each case the expectations are exactly balanced at a £15 gain. The first choice is in terms of 'gains', and so there may be a tendency for people to be risk-averse and prefer the 'safe' option of the additional £10. The second choice is in terms of 'losses', and so people may be more 'risk-seeking' and choose the 'risky' option, hoping that they will not need to return anything.

When the stakes are large

Small risks are one thing, but attitudes can change dramatically when there are very large amounts of money at stake. Consider the following choice.

Imagine you had to choose between

A   receiving 12p

B   having one of 1 000 000 tickets in a lottery with a prize of £100 000.

Which would you choose, A or B?

What is the expected gain of each option? The lottery option has an expected gain of 10p, and so receiving 12p seems the better choice, but many people may choose the 'risky' option as it gives a small chance of a life-changing event. People can be risk-seeking for *large* gains – this is why people buy lottery tickets.

Now consider the following (rather fictitious) choice.

Imagine you had to choose between

A   paying an insurance premium of £1.10

B   having a 1 in 1000 chance of losing property worth £1000.

Which would you choose, A or B?

What is the expected gain of each option? The risk-taking option, losing the property, has an expected loss of £1, and so might seem the better option, but many people will choose the 'safe' option of paying the insurance premium, as it protects against an unaffordable loss. People can be 'risk-averse' for *large* losses, which is why they buy insurance.

The preceding exercises may, if the students respond according to psychological research, reveal the pattern shown in Table 30.1.

	When dealing with potential gains	When dealing with potential losses
Small amounts of money at stake	Risk-averse – 'a bird in the hand is worth two in the bush'	Risk-seeking – 'in for a penny, in for a pound'
Large amounts of money at stake	Risk-seeking – buy lottery tickets	Risk-averse – buy insurance

Table 30.1 Common attitudes when dealing with gains and losses

In general, people have:

- a tendency to be risk-averse for small gains and large losses
- a tendency to be risk-seeking for large gains and small losses.

How we make such choices is complex, and there is no right answer, but having to decide on these various options should show students that dealing with probability and uncertainty in real life is more than just mathematics.

Deal or No Deal

This TV show has contestants deciding between a sequence of certain gains (offered by telephone from an anonymous 'banker') or a chance of a large (or small) gain from a remaining set of boxes. It has been suggested as a class activity, but requires either rather complex equipment [2] or access to an online game [3]. If it is currently on TV, it could be a topic for class discussion, covering issues such as the following.

- Are the contestants risk-averse or risk-seeking?
- How does the banker's offer change during the play, compared to the expected gain from the boxes?

The full rules [4] and also the possible strategy of the banker [5] are discussed on Wikipedia.

30.3 Further reading and resources

NRICH has a detailed list of questions to explore attitudes to risk and reward [6]. These issues are fully discussed in *Thinking, Fast and Slow* by Daniel Kahneman [7].

30.4 References

1   Wikipedia. *Framing effect (psychology)* [Internet]. [cited 2015 Oct 29]. Available from: https://en.wikipedia.org/wiki/Framing_effect_ (psychology)

2   DeRosa T. *Lesson Idea: Probability using Deal or No Deal* [Internet]. [cited 2015 Oct 29]. Available from: http://www.teachforever. com/2008/02/lesson-idea-probability-using-deal-or.html

3   Baker A, Bittner T, Makrigeorgis C, Johnson G, Haefner J. *Teaching Prospect Theory with the Deal or No Deal Game Show*. Teaching Statistics. 2010 Sep 1;32(3):81–7.

4   Wikipedia. *Deal or No Deal (UK game show)* [Internet]. [cited 2015 Oct 29]. Available from: https://en.wikipedia.org/wiki/Deal_or_ No_Deal_%28UK_game_show%29

5   Wikipedia. *Deal or No Deal* [Internet]. [cited 2015 Oct 29]. Available from: https://en.wikipedia.org/wiki/Deal_or_No_Deal

6   NRICH. *Discussing Risk and Reward* [Internet]. [cited 2015 Oct 29]. Available from: https://nrich.maths.org/6955

7   Kahneman D. *Thinking, Fast and Slow*. Farrar, Straus and Giroux; 2011. p.511

Happy birthday to you, and you, and …

31.1 Summary

A classic 'paradox' in probability is the so-called *birthday problem*: there need only be 23 people in a room for it to be more likely than not that at least two have a matching birthday (where 'birthday' is in terms of day and month, not year). Of course this is not a paradox at all, but just a particularly unintuitive result. The topic is a good one for classroom activity, and we show how it can be adapted to use birthdays of 'friends' on social media and to cover matches other than birthdays.

The birthday problem is a special case of the remarkable regularity of coincidences, and the mathematics can be developed in more generality than simply dealing with birthdays. We show a very useful approximation which makes it straightforward to calculate the number of individuals necessary to have a 50% or 95% chance of a match on any characteristic.

31.2 Possible classroom activities

Matching birthdays in class

If the class is larger than 23, then the odds are on your side that there will be at least one pair with matching birthdays, although the students are likely to be aware of this already. (In this situation twins count as one person!) Birthdays of a parent/sibling/other relation could be used as further rounds that should lead to more surprises.

With a large class, matches within rows/tables/groups can be explored using Table 31.2 towards the end of this chapter, where 'match' can be defined more loosely, say within 3 days.

Other questions that can be asked include:

- What about leap years – was anybody born on February 29th, or does anyone know someone who was?

- Are all days equally likely? In fact more babies are born in September, with 9% more than, say, in April. Lack of uniformity across the year very slightly increases the chance of a match. (Warning: possible reasons for the excess births in September include the influence of Christmas on parental activity, and parents planning to take advantage of the undoubted benefits of their child being older in the class, but this topic could lead to extensive and rather irrelevant discussion.)

Matching birthdays in sport

The birthday problem is often put in the context of football matches, since there are 23 people on the pitch – two teams of 11 and the referee.

> We would expect that in half of all football matches, two people on the pitch share a birthday. This could be explored, since birthdays of players are easily available online, although it would mean finding the birthday of the referee too.

Alternatively, the birthdays of whole squads can be explored, either in international competitions or in home teams. For example, out of 32 players in the Manchester United First Team Squad in November 2015, there was one pair with the same birthday: Chris Smalling and Marouane Fallaini were both born on 22nd November. In the 2014 World Cup there were 32 teams, and each team had a squad of 23 players. And, amazingly, in exactly half of the squads (16) there was a matching birthday – see BBC News [1].

Matching birthdays on Facebook

It is possible to use Facebook as a source of coincident birthdays, since birthdays of 'friends' are recorded under *Events*, although it may not be appropriate to explore Facebook in a class activity. Students can count the number of days on which there are matches. It will become clear that the chance of a match is dependent on how many friends you have, and students with a lot of 'friends' will have many coincident birthdays. Russell discusses using Facebook to illustrate the birthday problem [2].

Matching numbers

A surprising number of matches is not restricted to birthdays. If 20 people choose integers between 1 and 100 at random, for example, there is an 87% chance that at least two of them will choose the same number. This can make an activity that can be adapted to a class of any size.

> If there are n students in a class, challenge them to each choose a number between 1 and $\left(\frac{n}{2}\right)^2$ that is different from everyone else's number: they can, for example, simultaneously hold up a piece of paper with their number on it. If they truly choose at random, there is an 87% chance that at least two will choose the same number, and they will have lost the challenge. For example, in a class of 30, students would choose a number between 1 and $15^2 = 225$, and in almost every attempt at least two will choose the same number.

In fact the chances are far higher than 87%, since people find it difficult to choose a random number, and there will be an excess of numbers ending in 7, and few ending in 5 or 0, as well as an excess of extreme choices such as 1 or 225. Even when it is explained that the aim is to choose unique numbers, and this is best done by choosing at random, they will still find this very challenging. The teacher can keep on winning, even when the permissible range is increased.

The proof of this general result is given at the end of this chapter.

Counting handshakes

Some intuition for these 'paradoxes' can be gained through thinking of the number of ways in which matches can occur, which means counting the number of distinct pairings in the room. This can be explored by counting handshakes.

First get two students to shake hands – that is one handshake. Then get three students to all shake hands with each other – this will produce three handshakes. For four students there are six handshakes. What is the pattern? Table 31.1 can be used to record the results, and eventually the general formula of n people requiring $\frac{n(n-1)}{2}$ handshakes should be obtained.

Number of people in room	Number of handshakes required so that everyone has shaken hands with everyone else
2	1
3	3
4	6
5	10
...	...
n	$\frac{n(n-1)}{2}$

Table 31.1 Recording the results of handshakes

The general formula can also be found by direct reasoning: each of n people will need to shake hands with $(n-1)$ others, which is $n(n-1)$ handshakes – but this has counted each pairing twice, so we need to divide this total by 2. Alternatively, for those who have covered permutations and combinations, it is simply the number of ways of choosing 2 from n, or $^{n}C_{2} = \frac{n!}{(n-2)!2!} = \frac{n(n-1)}{2}$.

If there are 23 people in the room, and if they all shook hands with each other, there would be $\frac{23 \times 22}{2}$ handshakes, making 253 pairs of people shaking hands. There are therefore 253 possible pairs of people whose birthdays may match. The probability of any particular pair having a matching birthday is $\frac{1}{365}$, so the expected number of matches is $253 \times \frac{1}{365} = 0.69$, which is fairly high, so we should already be suspecting that it is not such a surprising occurrence (note that the expected number is equal to 'the probability of a match, times the number of possible matches' – the matches do not need to be independent for this to be true).

As we've seen in many examples, when dealing with events such as 'at least one match', it is usually easier to work out the chance of there being *not* a match. In matching problems it is best to think of all the people lined up and announcing their birthdays one by one. The first can be anything; the second must be different from the first, which has probability $\frac{364}{365}$; the third must be different from the first two, which has probability $\frac{363}{365}$, and so on, so that the probability that 23 people *all* have different birthdays is $\frac{364}{365} \times \frac{363}{365} \times \ldots \times \frac{343}{365} = 0.49$, so the probability there is at least one match is 0.51. This can be represented by the top branch of a probability tree (see Figure 31.1).

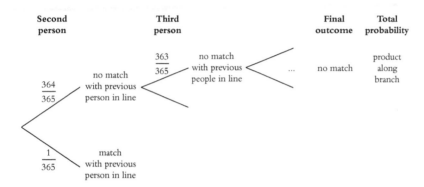

Figure 31.1 Probability tree for the birthday problem

In general, the chance of there being at least one match in n people is equal to $1 - \left(\frac{364}{365} \times \frac{363}{365} \times \ldots \times \frac{366-n}{365}\right)$.

The chance of someone out of n people sharing *your* birthday is

$$1 - \text{all } n \text{ people have a different birthday from yours} = 1 - \left(\frac{364}{365}\right)^n.$$

These probabilities are plotted in Figure 31.2.

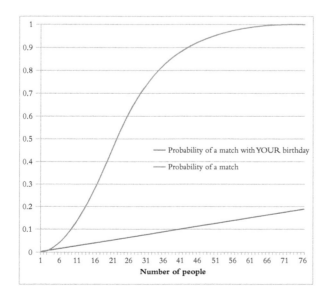

Figure 31.2 Graph of probability of birthday matches

There are useful approximations for such calculations.

Suppose the chance of any pair of people sharing a characteristic is $\frac{1}{K}$, where, for example, $K = 365$ when matching birthdays, or $K = 100$ when each person is asked to pick a random number between 1 and 100. Then for a 50% chance of a match, we need around $1.2\sqrt{K}$ people. For example, for birthdays the number needed for a 50% chance of a match is estimated to be $1.2\sqrt{365} = 23$ (exactly the right answer). For a 95% chance of a match, we need around $2.5\sqrt{K}$ people, so $2.5\sqrt{365} = 48$ people are needed for a 95% chance of at least two sharing a birthday. This helpful approximation is taken from Diaconis and Mosteller [3], a famous and utterly wonderful academic paper on coincidences, and the proof is given below.

These approximations mean it is straightforward to assess, for any size of group, what degree of matching is likely to occur (see Table 31.2).

Gap between birthdays	Probability 1 in K of 2 random people 'matching'	Number needed for chance of match to equal 50%	Number needed for chance of match to equal 95%
	K	$1.2\sqrt{K}$	$2.5\sqrt{K}$
Same day	365	23	48
Within 1 day	122	13	28
Within 3 days	52	9	18
Within 1 week	24	6	12
Within 2 weeks	13	4	9

Table 31.2 Numbers of people required for 50% and 95% chances of a close birthday match

For 9 people, for example, there is a 95% chance that two people will have a birthday within two weeks of each other, and a 50% chance within 3 days. The results in this table can be used to amaze and impress groups of people (and even win imaginary bets with them, if you are happy relying on a 95% chance of winning).

We can prove these approximations by generalising from the formula given above for birthdays. The probability $P(n, K)$ that there are *no* matches in a group of *n* people, when the probability of a match between a random pair is $\frac{1}{K}$, is equal to

$$P(n, K) = \left(1 - \frac{1}{K}\right) \times \left(1 - \frac{2}{K}\right) \times \left(1 - \frac{3}{K}\right) \times \dots \times \left(1 - \frac{(n-1)}{K}\right).$$

Provided *n* is reasonably small compared to *K*, then $\left(1 - \frac{t}{K}\right) \approx e^{-\frac{t}{K}}$ for $t = 1, 2, \dots, n - 1$, and so the probability that there are no matches is approximately

$$P(n, K) \approx e^{-\frac{[1+2+\dots+(n-1)]}{K}} = e^{-\frac{n(n-1)}{2K}}$$

using the fact that the sum of the integers from 1 to $n - 1$ is $\frac{n(n-1)}{2}$. For larger n this can be approximated by

$$P(n, K) \approx e^{-\frac{n^2}{2K}},$$

which gives $n = \sqrt{2K \ln\left(\dfrac{1}{P(n, K)}\right)}$.

Plugging in $P(n, K) = 0.5$ and 0.05, corresponding to a 50% and 95% chance of a match, gives the necessary approximations of $n = 1.2\sqrt{K}$ and $2.5\sqrt{K}$ respectively.

This also proves the earlier result on matching numbers: if you have n students in a class, and they choose random numbers between 1 and $K = \left(\frac{n}{2}\right)^2$, then the chance of no matches is $P(n, K) \approx e^{-\frac{n^2}{2K}} = e^{-2} = 0.13$. So the chance of a match is 87%.

31.3 Further reading and resources

A spreadsheet for these calculations is provided on the website.

Wikipedia has a detailed exposition of the birthday problem [4] – see also a detailed article by Borja and Haigh that covers non-uniformity of birthdays through the year [5]. NRICH has an excellent animation, *Same Number!*, which shows matching random numbers [6], while *What are the chances?* on the Understanding Uncertainty website derives the chances of picking the same number [7].

31.4 References

1   Fletcher J. *The birthday paradox at the World Cup* [Internet]. BBC News [cited 2015 Oct 16]. Available from: http://www.bbc.com/news/magazine-27835311

2   Russell M. *Pigeons, facebook and the birthday problem.* Teaching Statistics. 2013 Mar 1;35(1):26–8.

3   Diaconis P, Mosteller F. *Methods for Studying Coincidences.* Journal of the American Statistical Association. 1989;84(408):853–61.

4   Wikipedia. *Birthday problem* [Internet]. [cited 2015 Oct 16]. Available from: https://en.wikipedia.org/wiki/Birthday_problem

5   Borja MC, Haigh J. *The birthday problem.* Significance. 2007 Sep 1;4(3):124–7.

6   NRICH. *Same Number!* [Internet]. [cited 2015 Oct 16]. Available from: http://nrich.maths.org/7221

7   Understanding Uncertainty. *What are the chances?* [Internet]. [cited 2015 Oct 16]. Available from: http://understandinguncertainty.org/node/153

Chapter 32

How long do I have to wait?

32.1 Summary

This section considers how long we have to wait for events to happen, assuming a series of independent possibilities, such as how long we wait for a Head to occur when flipping coins. In particular we look at the expected or average time (the terms are used interchangeably). Note that this is essentially a discussion of the geometric distribution, although we do not address that explicitly.

This topic invites classroom participation, and yet involves some subtle mathematics. There is also a delightful scam perpetrated by illusionist Derren Brown.

32.2 Possible classroom activities

Last one standing

Each student stands up and everyone flips a coin at the same time. Everyone who gets a Head sits down. How many flips will it take until the whole class has sat down? In a class of 256, how many flips would you expect before everyone was sitting down?

As a class of students flipping coins while standing up is a recipe for complete chaos, this activity can also be done as a fun 'mind-reading' experiment. One person at the front flips a coin, while the rest of the class try to mind-read. They should put their hands on their heads if they think it is a Head, or on their hips if they think it is a Tail. Those who are wrong sit down. Unfortunately people tend to do the same thing as the group around them, and so the guesses will not be independent, and the class tends to sit down quicker than expected. For this reason, it can work better if students do the mind-reading with their eyes closed! Someone should make a string of correct guesses, and the class can discuss whether they really are a mind-reader.

We start by considering a class of 32. We expect that half the people will sit down at each flip, so 16 will be left standing after the first flip, then 8 after the second flip, 4 after three flips, 2 after four flips, 1 after five flips. So we expect around 5 or 6 flips before everyone has sat down, but of course in practice it may work out differently from this. In a class of 256, which is 2^8, we might expect 8 or 9 flips before everyone has sat down, as a few people might flip 7 or 8 Tails in a row, or guess this number of flips correctly.

NRICH's *Last One Standing* includes an animation to test how long it takes different size classes to all sit down [1].

Average wait for the first Head?

It seems intuitive that if an event, say throwing a six with a fair die, has a 1 in 6 chance of occurring, then on average the events should happen every 6 throws. Slightly less intuitive is the idea that if I start throwing a fair die, the average time until the first six is 6 throws. Proving this fact is also not straightforward.

> Consider flipping a fair coin. What is the average number of flips until the first Head?
>
> Get the class to try it 5 times each, using coins or spinners. Tally up the results: the most common 'waiting time' should be 1, and then the distribution of waiting times should drop down, roughly halving at each throw, although a few people may have flipped 5 or more Tails before they got a Head. Work out the average wait for a Head – it should be around 2. This seems intuitive, but can anyone prove it mathematically? This is a nice challenge.

The theory is shown in Figure 32.1.

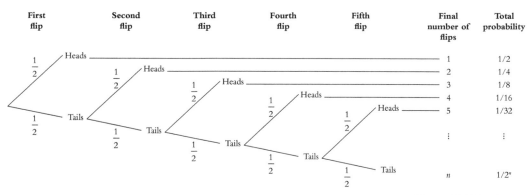

Figure 32.1 How long do you have to wait for a Head?

It should be clear from Figure 32.1 that the expected number of flips before the first Head, denoted E, is given by the equation

$$E = \left(1 \times \frac{1}{2}\right) + \left(2 \times \frac{1}{4}\right) + \left(3 \times \frac{1}{8}\right) + \left(4 \times \frac{1}{16}\right) + \left(5 \times \frac{1}{32}\right) + \cdots + \left(n \times \frac{1}{2^n}\right) + \cdots$$

This is not a trivial series to sum. The trick is to see that it can be written as

$$E = \left(\frac{1}{2} + \frac{1}{4} + \frac{1}{8} + \cdots + \frac{1}{2^n} + \cdots\right) + \left(\frac{1}{4} + \frac{1}{8} + \cdots + \frac{1}{2^n} + \cdots\right)$$
$$+ \left(\frac{1}{8} + \cdots + \frac{1}{2^n} + \cdots\right) + \cdots$$

and so, making use of the identity $1 = \frac{1}{2} + \frac{1}{4} + \frac{1}{8} + \frac{1}{2^n} + \cdots$, we get

$$E = 1 + \left(1 - \frac{1}{2}\right) + \left(1 - \frac{1}{2} - \frac{1}{4}\right) + \cdots = 1 + \frac{1}{2} + \frac{1}{4} + \cdots = 2$$

We will see a simpler, algebraic solution below in *How long to pass the test?*

> The average number of flips before both a Head and a Tail have turned up is 3. Why?

The first flip can be anything, and then we must wait for the other side to turn up. We have seen this has an expected (average) waiting time of 2 flips, so the total waiting time until both sides have appeared is 3.

How long to pass the test?

> Suppose there is probability p of someone passing their driving test, and each test is independent of other tests. How many times do they expect to have to take the test before they pass?

This can be solved using the 'summing series' method above, but it is much more neatly solved as a recursive, algebraic method. Suppose the expected number of attempts is E. If you pass the exam, you took 1 attempt; if you fail, the process starts all over again – as it has no memory - and you still have an expected E attempts in the future to pass, which means your overall expected number of attempts is now $E + 1$ (Figure 32.2).

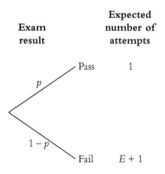

Figure 32.2 How many attempts will it take to pass the driving test?

So E obeys the identity $E = p + (1 - p)(E + 1)$. Solving this identity gives $E = \frac{1}{p}$. This is the required answer – the expected number of attempts until 'success' is just the inverse of the probability of success at each attempt: when trying to flip a Head, $p = \frac{1}{2}$ and so $E = 2$, the result we found by the brute-force summing-series approach.

The Derren Brown scam

Illusionist Derren Brown did a fine 'trick' in which he was filmed flipping a coin and it came up Heads 10 times in a row. This was a genuine piece of film, although it was revealed later in the programme that Brown had flipped the coin for hours, being filmed all the time, until he managed the feat – then only that tiny piece of film was shown. An internet search should reveal a YouTube video of this feat.

What is the probability of getting 10 Heads in a row when flipping a fair coin 10 times? What is the expected number of flips needed before getting 10 Heads in a row?

The chance of getting 10 Heads in a row, in a single attempt, is $\frac{1}{2^{10}} = \frac{1}{1024}$.

Any set of flips can be thought of as a series of 'sequences of Heads' terminated by a Tail – the sequence is of length 0 if another Tail is then flipped immediately. Now consider the expected number of 'sequences of Heads' necessary to get a run of 10, where the chance of a particular sequence consisting of 10 Heads is $\frac{1}{1024}$. By the general result shown for passing the driving test, $E = \frac{1}{p}$, and so we expect to need 1024 sequences. Each sequence must terminate in a Tail, but the average number of flips before the first Tail is 2. So the average length of a sequence of Heads, followed by its concluding Tail, is 2. We would therefore expect Derren Brown to need $2 \times 1024 = 2048 = 2^{11}$ flips (the exact expectation is 2046 flips – see below).

Say it takes around 4 seconds to flip a coin and observe the result, when working slowly and steadily, so there are 15 flips a minute. This means Brown expected to take $\frac{2048}{15} = 136$ minutes, or just over 2 hours to be successful. On the programme he said he filmed for 9 hours, which if true meant he was very unlucky. James Grime ('singingbanana') was luckier when he repeated the 'trick', and took only an hour – this can be seen on YouTube [2, 3].

It turns out that the exact formula for the expected number of flips before n Heads in a row is $E = 2(2^n - 1)$, which for large n is close to $2^{(n+1)}$ [4]. Exact values are given in Table 32.1.

Number of Heads in a row	Expected number of flips needed to achieve this
1	2
2	6
3	14
4	30
5	62
6	126
7	254
8	510
9	1022
10	2046

Table 32.1 Expected number of flips needed to observe different-length runs of Heads

NRICH has a detailed set of teachers' resources on this problem, including simulations that show there is around a 90% chance of completing the task by 5000 flips [5], taking perhaps 5 hours.

How long to get the set?

Each student has a fair die or equivalent spinner (like that used in Chapter 20), and has to count how many throws/spins are needed until they get all the faces at least once. There is a prize for the person who takes the longest and shortest time (no cheating).

It is clear that the minimum is 6 (which someone might achieve, although this is unlikely), and then there will be a long tail of results, with some people taking many throws/spins to get all the outcomes at least once. The average is likely to be around 15 throws/spins. Why?

The first throw can be anything. There are then are 5 numbers left, so the chance of getting a 'new' number on any throw is $\frac{5}{6}$, and the expected number of throws before getting a new number is $\frac{6}{5}$. Similarly, the chance of getting a 'new' number on the next throw is $\frac{4}{6}$, and so the expected number of throws before getting a third different number is $\frac{6}{4}$, and so on. The total expected number of throws is therefore $1 + \frac{6}{5} + \frac{6}{4} + \frac{6}{3} + \frac{6}{2} + 6 = 14.7$. So the observed average should be around 15, although there will be a *lot* of variation.

At the time of the World Cup or other major competitions, many people buy packets of football stickers to try and complete an album. In the 2014 World Cup, there were 640 stickers, so how long, on average, would it take to fill an album with all 640? What assumptions are being made, and are they realistic?

Assume that each sticker bought is like spinning a spinner with numbers 1 to 640 on it. Then, according to the reasoning given above, the expected number of stickers that needs to be bought to get all 640 is $1 + \frac{640}{639} + \frac{640}{638} + ... + 640$ which can, say using a spreadsheet, be evaluated to give 4505.3 stickers. This will cost a lot! The sum can also be approximated by $640(\ln(640) + 0.6) = 4520$. (For the mathematical detail of this approximation, see *Plus* magazine [6].)

In practice things are much more complicated as people can swap and buy cards online. We are also assuming that the cards in a packet are random, which they may not be. This whole area is known as the 'coupon collector's problem' [7]. Gerke [8] discusses this in detail.

32.3 Further reading and resources

A classic idea in probability is that if a large number of monkeys keep on typing at random, eventually they will produce the complete works of Shakespeare. Spiegelhalter and Smith [9] describe a computer program that randomly generates letters, and calculate that for a 90% chance of producing a 17 letter sequence that occurs anywhere in Shakespeare (such as 'to be or not to be'), the program would have to run for 6.6 billion years, around half the time since the Big Bang.

32.4 References

1   NRICH. *Last One Standing* [Internet]. [cited 2015 Oct 29]. Available from: http://nrich.maths.org/7220

2   singingbanana. *Flipping 10 heads in a row* [Internet]. YouTube. [cited 2015 Oct 30]. Available from: http://www.YouTube.com/watch?v=_K585ODq0a0

3   singingbanana. *Flipping 10 heads in a row: Full video* [Internet]. YouTube. [cited 2015 Oct 30]. Available from: https://www.YouTube.com/watch?v=rwvIGNXY21Y

4   Spivey M. *The Expected Number of Flips for a Coin to Achieve n Consecutive Heads* [Internet]. A Narrow Margin. [cited 2015 Oct 30]. Available from: https://mikespivey.wordpress.com/2012/02/09/consecutiveheads/

5   NRICH. *The Derren Brown Coin Flipping Scam* [Internet]. [cited 2015 Oct 30]. Available from: https://nrich.maths.org/6954

6   Barrow J. *Outer space: A collector's piece* [Internet]. [cited 2015 Oct 30]. Available from: https://plus.maths.org/content/os/issue37/outerspace/index

7   Wikipedia. *Coupon collector's problem* [Internet]. [cited 2015 Oct 30]. Available from: https://en.wikipedia.org/wiki/Coupon_collector%27s_problem

8   Gerke O. *How much is it going to cost me to complete a collection of football trading cards?* Teaching Statistics. 2013 Jun 1;35(2):89-93.

9   Spiegelhalter D, Smith O. *Understanding uncertainty: Infinite monkey business* [Internet]. [cited 2015 Oct 30]. Available from: https://plus.maths.org/content/infinite-monkey-business

Do you know what you don't know?

33.1 Summary

Probability is usually identified with ideas of chance or randomness – unavoidable unpredictability about the future. The idea of this exercise is to encourage numerical expressions of uncertainty about *facts* – to assess the 'probability that I am right'. Students' skill in doing this is assessed using a quiz – the higher the probability they give to a correct answer, the better they score.

33.2 Possible classroom activities

Epistemic uncertainty

Take a fair coin, and ask the class the probability it will come up Heads if you flip it. They should, with luck, say '$\frac{1}{2}$'. Then flip the coin, but cover it up before you see the result. Now ask – what is the probability that it is a Head? Eventually someone may, very grudgingly, say '$\frac{1}{2}$', perhaps wondering what you are getting at.

There is a reluctance to give a probability as there is no longer any randomness – the coin is either Heads or Tails. But neither you nor the class know which, so if we take a 'subjective' interpretation of probability, then you can still say that the probability is $\frac{1}{2}$, since this expresses your uncertainty about what you will see when you uncover the coin.

The personal nature of such a probability is very clearly shown if you now look at the coin, but without showing the class. *Your* probability is now 1 for whatever the coin shows, whereas the probability for the class stays at $\frac{1}{2}$. This should make clear that such a probability is not a property of the coin alone, but of your knowledge about the coin.

Before the coin is flipped, when we *can't* know the answer, our uncertainty is known as **aleatory** – also known as chance, randomness, unpredictability. Afterwards, when we *don't* know the answer, our uncertainty is known as **epistemic** – essentially lack of knowledge, or ignorance.

A quiz

Give the following instructions [*documents are provided on our website*].

You are going to hear a set of questions with two possible answers: A or B. You need to think which answer you prefer, but then say how confident you are that you are right.

- If you are *certain* that you are right, give yourself a 'confidence' of 10 (corresponding to a probability of 1, or 100%).

- If you have *no idea* whether the answer is A or B, give yourself a 'confidence' of 5 (corresponding to a probability of 0.5, or 50%).

- If you think that you know the answer, but are not absolutely sure, then give yourself a 'confidence' of 6, 7, 8 or 9. So if you are fairly confident, you might give your answer a confidence of 8.

You will then find out the true answer, and work out your score from Table 33.1.

Your 'confidence' in your answer	5	6	7	8	9	10
Score if you are *right*	0	9	16	21	24	25
Score if you are *wrong*	0	−11	−24	−39	−56	−75

Table 33.1 Scoring answers relative to your confidence in them

Then use Table 33.2 to keep track of your scores, and to find your total score for the ten questions.

Question	My preferred answer	My confidence in my answer	The true answer	Right or wrong?	Score
1					
...					
10					
Total score					

Table 33.2 Score sheet for quiz

Typical questions might be of the following type, possibly updated and adapted to contemporary circumstances.

- Who has the most followers on Twitter, A) Justin Bieber or B) Rihanna?
  A (A – 49 million vs B – 34 million)

- Who is older, A) Prince William or B) Kate (Duchess of Cambridge)?
  B (A – born 21/6/82 vs B – born 9/1/82)

- In the IMDB rankings (November 2015), which film comes higher, A) WALL.E or B) Toy Story 3?
  A (A – 61st vs B – 79th)

- Which is larger, A) Belgium or B) Switzerland?
  B (A – 30 000 sq km vs B – 41 000 sq km)

- Which is bigger, A) Venus or B) Earth?
  B (A – 6051 km vs B – 6371 km radius)

What kind of behaviour is rewarded – and penalised? The aim is to try to reward both 'knowing a lot' and also 'knowing what you know'. People who know a lot, or are very lucky, will score highly, while people who do not know much, but are aware of this and so are cautious, will end up scoring around zero. The strongest penalty is reserved for those who don't know, but think they do: a lethal mixture of cockiness and ignorance that nobody wants in an advisor or predictor (anecdotally, young males appear particularly prone to this behaviour).

This exercise can easily be extended to TV talent contests, celebrity competitions, *The Great British Bake Off*, *Strictly Come Dancing*, and so on. Each student chooses who they think will get eliminated in the next round, gives their confidence, and then gets scored after the result has been announced. Scores can be accumulated over the series. Sporting events are ideal. Instead of students just saying who they think will win, they have to give their confidence and can be scored over a competition, say the World Cup. A draw means a score of 0.

It has also been proposed that similar methods should be used instead of standard multiple-choice questions in which a single answer is permitted: so-called 'certainty-based marking' [1].

The scoring rule is non-linear and asymmetric in gains and losses – a confident opinion that turns out to be incorrect loses far more than it would gain if it is right, rather like a nasty teacher who punishes failure harshly but only grudgingly rewards success. However, this property is not arbitrary: the rule is carefully designed to reward honest expression of opinion and to discourage over-stating one's confidence.

What is the formula behind the scores? (Hint: subtract each score from the maximum, 25, and observe the pattern: 0, 1, 4, 9, 16, and so on).

Let X be the number given to the correct answer and $10 - X$ the number assigned to an incorrect answer, so that supposing you say '8', if you are right then $X = 8$, and if you are wrong then $X = 2$. The formula for your score is then $25 - (10 - X)^2$.

From your maximum of 25 per question, you essentially lose the *square* of your error. This scoring rule is a transformed version of the Brier scoring rule [2] developed to train and evaluate the probabilistic predictions of weather forecasters. It is an example of what is known as a *proper scoring rule*, which encourages people to honestly express their beliefs, in the sense that their expected score, calculated with respect to their 'true' probability, is maximised by choosing their level of confidence to match that probability.

To show that this scoring rule encourages honesty, suppose my honest probability for B was 70%, and so I chose 7 as my confidence level. Then I judge that I have a 70% probability of gaining 16, and a 30% probability of losing 24, and so my expected score is $(0.7 \times 16) - (0.3 \times 24) = 4.0$. If I was arrogant, on the other hand, choosing to exaggerate by claiming 100% confidence, then my expected score is $(0.7 \times 25) - (0.3 \times 75) = -5.0$, which is less than if I had chosen to express my true opinion. So although I could be lucky in this instance, on average it will pay me to express my uncertainty honestly.

Suppose instead we used the following rule in Table 33.3, which is linear and symmetric, and may superficially seem reasonable as it essentially penalises by the distance from the correct answer.

Your 'confidence' in your answer	5	6	7	8	9	10
Score if you are *right*	0	5	10	15	20	25
Score if you are *wrong*	0	−5	−10	−15	−20	−25

Table 33.3 Alternative (inappropriate) 'linear' scoring rule for quiz

Then my expected score by being honest is $(0.7 \times 10) - (0.3 \times 10) = 4.0$ as before, whereas my expected score if I exaggerate is $(0.7 \times 25) - (0.3 \times 25) = 10.0$. So this linear rule, although apparently reasonable, actually encourages people to lie about their uncertainty. Unfortunately this rule has been used in some studies.

33.3 Further reading and resources

Much of this material is taken from Understanding Uncertainty's *Do you know what you know?* [3], which features an interactive quiz with a large stock of questions.

33.4 References

1   LAPT. *Certainty-Based Marking* [Internet]. [cited 2015 Oct 17]. Available from: http://www.ucl.ac.uk/lapt/

2   Wikipedia. *Brier score* [Internet]. [cited 2015 Oct 17]. Available from: https://en.wikipedia.org/wiki/Brier_score

3   Understanding Uncertainty. *Do you know what you know?* [Internet]. [cited 2015 Oct 17]. Available from: http://understandinguncertainty. org/do-you-know-what-you-know

Index

Milton Keynes UK
Ingram Content Group UK Ltd.
UKHW050108171024
449665UK00005B/62

9 781316 605899